Problem Solvers

Edited by L. Marder
Senior Lecturer in Mathematics, University of Southampton

No. 4

Analytical Mechanics

Problem Solvers

Analytical Mechanics

D. F. LAWDEN, M.A., Sc.D., F.R.S.N.Z.
Professor of Mathematical Physics,
University of Aston in Birmingham

LONDON · GEORGE ALLEN & UNWIN LTD
RUSKIN HOUSE · MUSEUM STREET

First published in 1972

© George Allen & Unwin Ltd, 1972

ISBN 0 04 531004 1 *hardback*
0 04 531005 x *paperback*

Printed in Great Britain
in 10 on 12 pt 'Monophoto' Times Mathematics Series 569
by Page Bros. (Norwich) Ltd., Norwich

Contents

Preface

In writing this contribution to the Problem Solvers series, I have assumed that the reader has completed a first course in mechanics and is familiar with the techniques employed to solve problems relating to the motion of a single particle or a small number of interacting particles. In particular, I have supposed that his knowledge of the subject embraces inertial frames, Newton's laws, the velocity and acceleration vectors and their components, properties of idealized components of mechanical systems such as inextensible strings, perfectly rough planes, etc., and the definitions of work and energy. However, some of the problems included in Chapter 2 serve to recall these fundamental concepts to mind.

The prime purpose of the book is to illustrate principles and methods by which problems concerning the behaviour of systems of rigid bodies may be analysed. The calculation of the inertial characteristics of such systems is studied in Chapter 1. Principles of linear and angular momenta are applied in Chapter 2 to the calculation of plane motions and in Chapter 3 to the calculation of three dimensional motions. In Chapter 4, the methods associated with the names of Lagrange and Hamilton are used to provide alternative solutions of some of the earlier problems (the reader is thus able to compare their relative effectiveness) and are also applied to many new systems. The book concludes with a number of problems relating to oscillatory systems having two or three degrees of freedom.

The topics included are those which commonly form part of an honours degree course in applied mathematics and it is hoped, therefore, that a useful supplement to the available texts of analytical mechanics has been provided. The exercises at the end of the chapters form a fairly stern test of the student's ability to apply the techniques illustrated in the main body of the book.

In conclusion, it is a pleasure to acknowledge the assistance of my secretary, Mrs. Audrey Breakspear, who performed so efficiently the difficult task of typing the original manuscript.

D. F. L.

Chapter 1

Characteristics of Mass Distributions

1.1 Centres of Mass If a mechanical system comprises n particles of masses $m_i (i = 1, 2, \ldots, n)$ situated at points P_i having position vectors $\mathbf{r}_i$ with respect to an origin O, its centre of mass is the point G having position vector $\bar{\mathbf{r}}$, determined by the equation

$$\bar{\mathbf{r}} = \sum_{i=1}^{n} m_i \mathbf{r}_i / \sum_{i=1}^{n} m_i. \qquad (1.1)$$

If (x_i, y_i, z_i) are the coordinates of the particle m_i with respect to a rectangular cartesian coordinate frame $Oxyz$ and $(\bar{x}, \bar{y}, \bar{z})$ are the coordinates of G, then

$$\bar{x} = \sum m_i x_i / \sum m_i, \quad \text{etc.} \qquad (1.2)$$

In the limiting case when the system is regarded as a continuous distribution of matter over a region of space Γ, $\rho(x, y, z)$ being the density, then sums are replaced by integrals, thus:

$$\bar{x} = \int_\Gamma \rho x \, dv \Big/ \int_\Gamma \rho \, dv, \quad \text{etc.,} \qquad (1.3)$$

the integrals being volume integrals assuming that Γ is a three-dimensional region; if Γ is a surface, then ρ is taken to be the surface density and the integrals are evaluated over the surface; if Γ is a curve, ρ is the line density and the integrals are line integrals.

Problem 1.1 Two distinct distributions of matter have total masses M_1, M_2 and their centres of mass are at the points $\bar{\mathbf{r}}_1$, $\bar{\mathbf{r}}_2$. Find the centre of mass of the combined distribution.

Solution. Let $m_i (i = 1, 2, \ldots, n)$ be the particles of the first distribution at the points $\mathbf{r}_i$ and $m'_j (j = 1, 2, \ldots, n')$ be the particles of the second at the points $\mathbf{r}'_j$. Then

$$M_1 \bar{\mathbf{r}}_1 = \sum_i m_i \mathbf{r}_i, \qquad M_2 \bar{\mathbf{r}}_2 = \sum_j m'_j \mathbf{r}'_j.$$

If the centre of mass required is at $\bar{\mathbf{r}}$, then

$$(M_1 + M_2)\bar{\mathbf{r}} = \sum_i m_i \mathbf{r}_i + \sum_j m'_j \mathbf{r}'_j = M_1 \bar{\mathbf{r}}_1 + M_2 \bar{\mathbf{r}}_2. \qquad (1.4)$$

This equation determines $\bar{\mathbf{r}}$. $\qquad\qquad\qquad\qquad\qquad\qquad\qquad\qquad$ □

Problem 1.2 The density inside a sphere of radius a is proportional to the depth beneath the surface. Determine the centre of mass of a hemisphere.

Solution. Taking axes Ox, Oy in the plane face of the hemisphere, with O at its centre, it is obvious that the centre of mass lies on Oz; let $(0,0,\bar{z})$ be its coordinates. Introducing spherical polar coordinates (r, θ, ϕ) (Fig. 1.1), the density is given by

$$\rho = \lambda(a-r).$$

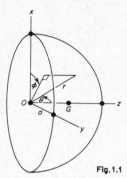

Fig.1.1

Employing equation (1.3) and expressing the volume integrals in spherical polars, it follows that

$$\bar{z} = \iiint \rho z . r^2 \sin\theta \, dr d\theta d\phi \div \iiint \rho . r^2 \sin\theta \, dr d\theta d\phi.$$

Since $z = r \cos\theta$, the first of these integrals reduces to the triple integral

$$\int_0^a dr \int_0^{\frac{1}{2}\pi} d\theta \int_0^{2\pi} \lambda(a-r)r^3 \sin\theta \cos\theta \, d\phi = \pi\lambda a^5/20.$$

The second integral is evaluated in the same way and yields the result $\pi\lambda a^4/6$. Thus $\bar{z} = \frac{3}{10}a$. ∎

Problem 1.3 The radii of the ends of a frustum of a uniform, solid, right circular cone are a and $2a$. Show that the distances of its mass centre from the ends are in the ratio $17:11$.

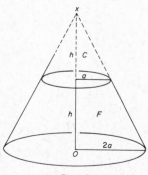

Fig.1.2

Solution. Suppose that a cone C of the same material and base radius a is placed on the frustum F (Fig. 1.2), thus generating a cone $F+C$ of base radius $2a$. If h is the height of F, then h is also the height of C. Taking the x-axis as indicated, the mass centres of F, C and $F+C$ lie on this axis and have x-coordinates $\bar{x}$, $5h/4$, $h/2$, respectively (the centre of mass of a uniform cone of height h is a distance $\frac{1}{4}h$ from the base). If ρ is the density of the material, the masses of F and C are $7\pi a^2 h\rho/3$, $\pi a^2 h\rho/3$, respectively. Employing equation (1.4) to give the x-coordinate of the centre of mass of $F+C$, we get

$$\frac{1}{2}h = \left(\frac{7}{3}\bar{x}+\frac{1}{3}\cdot\frac{5}{4}h\right)\Big/\left(\frac{7}{3}+\frac{1}{3}\right).$$

Whence $\bar{x} = 11h/28$, which is equivalent to the result stated. $\qquad\square$

1.2 Moments of Inertia

Let l be a straight line and let $p_i(i = 1, 2, \dots, n)$ be the perpendicular distances of particles $m_i(i = 1, 2, \dots, n)$ from l. Then, the *moment of inertia* of the system of particles about the axis l is I, where

$$I = \sum_{i=1}^{n} m_i p_i^2. \tag{1.5}$$

If m_i has coordinates (x_i, y_i, z_i) in a rectangular frame $Oxyz$, the moments of inertia of the system about the coordinate axes are therefore

$$A = \sum m_i(y_i^2 + z_i^2), \quad B = \sum m_i(z_i^2 + x_i^2), \quad C = \sum m_i(x_i^2 + y_i^2), \tag{1.6}$$

respectively. In the limiting case of a continuous distribution of matter having density $\rho(x, y, z)$ over a region Γ,

$$A = \int_{\Gamma} \rho(y^2 + z^2)\, dv, \quad \text{etc.} \tag{1.7}$$

If M is the total mass of a matter distribution and I is its moment of inertia about a certain axis l, the radius of gyration of the distribution about l is k, where

$$I = Mk^2. \tag{1.8}$$

Problem 1.4 Calculate the radius of gyration of the sphere described in Problem 1.2 about one of its diameters.

Solution. Taking the centre as origin and the diameter as z-axis, the moment of inertia is the volume integral

$$\int \rho(x^2 + y^2)\, dv,$$

calculated over the sphere. Introducing polar coordinates as in Problem 1.2, $x^2 + y^2 = r^2 \sin^2 \theta$ and we have accordingly to evaluate the triple integral

$$\int_0^a dr \int_0^\pi d\theta \int_0^{2\pi} \lambda(a-r)r^4 \sin^3\theta\, d\phi.$$

The result is $4\pi\lambda a^6/45$.

The mass of a hemisphere was found in Problem 1.1 to be $\pi\lambda a^4/6$. Hence for the sphere, $M = \pi\lambda a^4/3$. The radius of gyration now follows from equation (1.8): $k = 2a/\sqrt{15}$. ◻

1.3 Products of Inertia

Problem 1.5 Calculate the moment of inertia of a matter distribution about an axis L, having direction cosines (l, m, n) relative to a frame $Oxyz$, and passing through O.

Solution. Let m_i be a typical particle at $P(x_i, y_i, z_i)$. PN is perpendicular to the axis L (Fig. 1.3). Then

$$PN^2 = OP^2 - ON^2 = r_i^2 - (r_i \cos\alpha)^2.$$

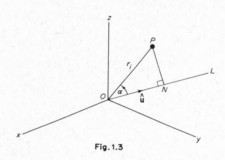

Fig. 1.3

But $r_i\cos\alpha = \mathbf{r}_i.\hat{\mathbf{u}} = lx_i + my_i + nz_i$, $\hat{\mathbf{u}} = (l, m, n)$ being the unit vector in the direction of L. Hence the required moment of inertia is

$$\sum m_i PN^2 = \sum m_i [x_i^2 + y_i^2 + z_i^2 - (lx_i + my_i + nz_i)^2],$$
$$= Al^2 + Bm^2 + Cn^2 - 2Fmn - 2Gnl - 2Hlm, \qquad (1.9)$$

since $l^2 + m^2 + n^2 = 1$; A, B, C have been defined in the previous section and

$$F = \sum m_i y_i z_i, \quad G = \sum m_i z_i x_i, \quad H = \sum m_i x_i y_i. \qquad (1.10)$$

F, G, H are called the *products of inertia* of the masses m_i with respect to the frame $Oxyz$. If the plane Oyz is a plane of symmetry for the matter distribution, then $G = H = 0$. If Ozx is a plane of symmetry, then $H = F = 0$ and if Oxy is a plane of symmetry, $F = G = 0$. ◻

For a continuous distribution, equations (1.10) are replaced by

$$F = \int \rho yz \, dv, \quad \text{etc.} \qquad (1.11)$$

Problem 1.6 A uniform plate is in the shape of a quadrant of a circle of radius a. If M is its mass, calculate its moment of inertia about an axis through the corner inclined at an angle α to one of the straight edges.

Solution. The moment of inertia of the plate about a straight edge is clearly one quarter of the moment of inertia of a complete circular plate about a diameter. The latter is well-known to be $\frac{1}{4}M'a^2$, where $M' = 4M$ is the mass of the circular plate. Thus, if we take the axes Ox, Oy along the straight edges, then

$$A = B = \tfrac{1}{4}Ma^2.$$

The product of inertia with respect to the axes Oxy is

$$H = \int \rho xy \, dA,$$

where dA is an element of area of the plate and ρ is its mass per unit area. Transforming to polar coordinates (r, θ), H can be expressed as a repeated integral thus:

$$H = \int_0^a dr \int_0^{\frac{1}{2}\pi} \rho r^3 \sin \theta \cos \theta \, d\theta$$
$$= \tfrac{1}{8}\rho a^4 = Ma^2/2\pi.$$

The direction cosines of the axis are $(\cos \alpha, \sin \alpha, 0)$. Substituting in (1.9), therefore, the required moment of inertia is

$$\frac{1}{4}Ma^2(\cos^2\alpha + \sin^2\alpha) - \frac{1}{\pi}Ma^2\sin \alpha \cos \alpha = Ma^2\left(\frac{1}{4} - \frac{1}{2\pi}\sin 2\alpha\right). \qquad \square$$

1.4 Theorems of Parallel and Perpendicular Axes

If I_G is the moment of inertia of a matter distribution about an axis through the centre of mass and I is the moment of inertia of the same matter distribution about a parallel axis distant d from the first axis, then

$$I = I_G + Md^2, \qquad (1.12)$$

where M is the total mass of the distribution. This is the *theorem of parallel axes for moments of inertia*.

Let GX, GY, GZ be rectangular axes through the centre of mass of a distribution and let Ox, Oy, Oz be parallel axes through any point O. Let $(\bar{x}, \bar{y}, \bar{z})$ be the coordinates of G with respect to the frame $Oxyz$. Then, if A_G, B_G, C_G are the moments of inertia of the distribution about the axes GX, GY, GZ respectively and A, B, C are the moments of inertia about the axes Ox, Oy, Oz, it follows from (1.12) that

$$\left. \begin{aligned} A &= A_G + M(\bar{y}^2 + \bar{z}^2), \\ B &= B_G + M(\bar{z}^2 + \bar{x}^2), \\ C &= C_G + M(\bar{x}^2 + \bar{y}^2). \end{aligned} \right\} \qquad (1.13)$$

5

Further, if F_G, G_G, H_G are the products of inertia relative to the frame $GXYZ$ and F, G, H are the corresponding products of inertia in the frame $Oxyz$, then it may be shown that

$$F = F_G + M\bar{y}\bar{z}, \quad G = G_G + M\bar{z}\bar{x}, \quad H = H_G + M\bar{x}\bar{y}. \quad (1.14)$$

This is the *theorem of parallel axes for products of inertia*.

If the axes Ox, Oy are taken in the plane of a thin plate (lamina) and Oz is perpendicular to the plate, then the moments of inertia of the plate about the axes are related by the equation

$$C = A + B. \quad (1.15)$$

This is the *theorem of perpendicular axes* for a lamina.

Problem 1.7 Ox, Oy, Oz lie along three concurrent edges of a uniform rectangular block of mass M. The lengths of these edges are a, b, c respectively. Calculate the products of inertia of the block in the frame $Oxyz$ and hence find its moment of inertia about a diagonal of a face with dimensions a and b.

Solution. Taking parallel axes through the centre G of the block, the moments of inertia about these axes are

$$A_G = \tfrac{1}{12}M(b^2 + c^2), \quad B_G = \tfrac{1}{12}M(c^2 + a^2), \quad C_G = \tfrac{1}{12}M(a^2 + b^2).$$

Since the block is symmetric with respect to these axes, the products of inertia all vanish in this frame. Employing the parallel axes theorems, the moments of inertia and products of inertia in the frame $Oxyz$ are now calculated to be ($\bar{x} = \tfrac{1}{2}a$, $\bar{y} = \tfrac{1}{2}b$, $\bar{z} = \tfrac{1}{2}c$)

$$A = \tfrac{1}{3}M(b^2 + c^2), \quad B = \tfrac{1}{3}M(c^2 + a^2), \quad C = \tfrac{1}{3}M(a^2 + b^2),$$
$$F = \tfrac{1}{4}Mbc, \quad G = \tfrac{1}{4}Mca, \quad H = \tfrac{1}{4}Mab.$$

The diagonal of the face lying in the xy-plane has direction cosines $(\lambda a, \lambda b, 0)$, where $\lambda = (a^2 + b^2)^{-\frac{1}{2}}$. Substituting in equation (1.9), the moment of inertia about this diagonal is now found to be

$$\frac{M(a^2b^2 + 2b^2c^2 + 2c^2a^2)}{6(a^2 + b^2)}. \qquad \square$$

1.5 Principal Axes of Inertia   Let $Oxyz$, $Ox'y'z'$ be two rectangular cartesian frames with a common origin. The transformation relating the coordinates of a point in the first frame to the coordinates of the same point in the second frame, can be written

$$\left. \begin{aligned} x' &= l_{x'x}x + l_{x'y}y + l_{x'z}z, \\ y' &= l_{y'x}x + l_{y'y}y + l_{z'z}z, \\ z' &= l_{z'x}x + l_{z'y}y + l_{z'z}z, \end{aligned} \right\} \quad (1.16)$$

where $l_{x'x}$ is the cosine of the angle between the axes Ox' and Ox, etc. Denoting the column matrix with elements (x, y, z) by $\mathbf{x}$ and the column matrix with elements (x', y', z') by $\mathbf{x}'$, this transformation can be expressed in matrix form thus:

$$\mathbf{x}' = \mathbf{l}\mathbf{x}, \tag{1.17}$$

where

$$\mathbf{l} = \begin{bmatrix} l_{x'x} & l_{x'y} & l_{x'z} \\ l_{y'x} & l_{y'y} & l_{y'z} \\ l_{z'x} & l_{z'y} & l_{z'z} \end{bmatrix}. \tag{1.18}$$

The elements in the first row of this matrix are the direction cosines of Ox' in the frame $Oxyz$, the second row are the direction cosines of Oy' and the third row are the direction cosines of Oz'. The matrix $\mathbf{l}$ is *orthogonal* and hence satisfies the condition

$$\mathbf{l}^{-1} = \mathbf{l}^{T}. \tag{1.19}$$

Let A, B, C, F, G, H be the moments and products of inertia relative to the frame $Oxyz$ and let A', etc. be the corresponding quantities in the 'dashed' frame. Form the matrix $\mathbf{J}$ thus:

$$\mathbf{J} = \begin{bmatrix} A & -H & -G \\ -H & B & -F \\ -G & -F & C \end{bmatrix}. \tag{1.20}$$

The matrix $\mathbf{J}'$ is constructed in the same way from the coefficients of inertia in the frame $Ox'y'z'$. Then

$$\mathbf{J}' = \mathbf{l}\mathbf{J}\mathbf{l}^{T}. \tag{1.21}$$

This equation, together with the parallel axes theorems of the previous section, enable us to calculate the coefficients of inertia in any rectangular frame when these coefficients are known in one such frame.

It is a well-known algebraic result that, given any symmetric matrix $\mathbf{J}$, it is possible to find an orthogonal matrix $\mathbf{l}$ such that the matrix $\mathbf{J}'$ given by equation (1.21) is diagonal. After transformation to such a frame $Ox^*y^*z^*$, the products of inertia F^*, G^*, H^* will all vanish. The axes Ox^*, Oy^*, Oz^* are then called *principal axes of inertia* and the moments of inertia about them, A^*, B^*, C^*, are the *principal moments of inertia*.

If (l, m, n) are the direction cosines of a line through O with respect to the principal axes, the moment of inertia of the distribution about this line is given by

$$I = A^*l^2 + B^*m^2 + C^*n^2. \tag{1.22}$$

This is the special form taken by equation (1.9) for principal axes.

Considerations of symmetry often permit the principal axes through a point to be recognized without calculation. If this is not the case, we determine the matrix l which transforms from given axes $Oxyz$ to principal axes through O, as follows. Solve the determinantal equation

$$|\mathbf{J} - \lambda \mathbf{I}| = 0 \qquad (1.23)$$

for λ ($\mathbf{I}$ is the unit matrix). Let $\lambda_1, \lambda_2, \lambda_3$ be the three roots (these are always real). Solve the three homogeneous equations

$$(\mathbf{J} - \lambda_1 \mathbf{I})\mathbf{x} = 0 \qquad (1.24)$$

for the ratios $x : y : z$ and choose values of x, y, z to satisfy the condition

$$x^2 + y^2 + z^2 = 1. \qquad (1.25)$$

These three values are $l_{x'x}, l_{x'y}, l_{x'z}$, (i.e. the direction cosines of Ox' in $Oxyz$) and hence give the first row of l. The second and third rows of l are found in the same manner after replacing λ_1 by λ_2 and λ_3 respectively. The principal moments of inertia are given by

$$A^* = \lambda_1, \qquad B^* = \lambda_2, \qquad C^* = \lambda_3. \qquad (1.26)$$

If $\lambda_1 = \lambda_2 \neq \lambda_3$, calculate the direction cosines of Oz' in $Oxyz$ from the root λ_3 and so write down the third row of l. The axes Ox', Oy' may be taken in any pair of perpendicular directions which are both perpendicular to Oz'.

If $\lambda_1 = \lambda_2 = \lambda_3$, all sets of perpendicular axes through O are principal axes.

Problem 1.8 Uniform rods OA, AB, BC, are each of mass M and length a and are mutually perpendicular. Determine the principal axes and moments of inertia at the centre of mass of this sytem.

Solution. Taking x-axis along OA, y-axis through O parallel to AB and z-axis parallel to BC, the centres of mass of the rods have coordinates $(\frac{1}{2}a, 0, 0)$, $(a, \frac{1}{2}a, 0)$, $(a, a, \frac{1}{2}a)$. Let $(\bar{x}, \bar{y}, \bar{z})$ be the centre of mass of the system. Then

$$3M\bar{x} = M(\tfrac{1}{2}a + a + a), \quad 3M\bar{y} = M(\tfrac{1}{2}a + a), \quad 3M\bar{z} = M \cdot \tfrac{1}{2}a.$$

Thus $\bar{x} = 5a/6$, $\bar{y} = a/2$, $\bar{z} = a/6$.

In the frame $Oxyz$, the moments of inertia about the axes of the whole system are easily found to be

$$A = \tfrac{5}{3}Ma^2, \qquad B = \tfrac{8}{3}Ma^2, \qquad C = \tfrac{11}{3}Ma^2.$$

The contributions of the rods OA and AB to the product of inertia F are zero, since every particle of these rods has zero z-coordinate. The contribution of the rod BC is

$$\int_0^a \rho a z \, dz = \tfrac{1}{2}\rho a^3 = \tfrac{1}{2}Ma^2,$$

8

ρ being the mass per unit length. Thus
$$F = \tfrac{1}{2}Ma^2.$$
Similar calculations yield
$$G = \tfrac{1}{2}Ma^2, \qquad H = \tfrac{3}{2}Ma^2.$$
We next apply the parallel axes theorems to give the moments of inertia and products of inertia with respect to a parallel frame through the centre of mass G. The results are
$$A_G = \tfrac{5}{6}Ma^2, \qquad B_G = \tfrac{1}{2}Ma^2, \qquad C_G = \tfrac{5}{6}Ma^2,$$
$$F_G = \tfrac{1}{4}Ma^2, \qquad G_G = \tfrac{1}{12}Ma^2, \qquad H_G = \tfrac{1}{4}Ma^2.$$
The determinantal equation (1.23) can now be constructed. It is
$$\begin{vmatrix} \tfrac{5}{6}-\mu & -\tfrac{1}{4} & -\tfrac{1}{12} \\ -\tfrac{1}{4} & \tfrac{1}{2}-\mu & -\tfrac{1}{4} \\ -\tfrac{1}{12} & -\tfrac{1}{4} & \tfrac{5}{6}-\mu \end{vmatrix} = 0,$$
where $\mu = \lambda/Ma^2$. The characteristic roots are $\mu = 1$, $\tfrac{11}{12}$ and $\tfrac{1}{4}$, implying that the principal moments of inertia at G are
$$A^* = Ma^2, \quad B^* = \tfrac{11}{12}Ma^2, \quad C^* = \tfrac{1}{4}Ma^2.$$
Taking $\lambda = Ma^2$, equations (1.24) take the form
$$-\frac{x}{6}-\frac{y}{4}-\frac{z}{12} = 0, \quad -\frac{x}{4}-\frac{y}{2}-\frac{z}{4} = 0, \quad -\frac{x}{12}-\frac{y}{4}-\frac{z}{6} = 0.$$
These have the solution $x:y:z = 1:-1:1$. To satisfy the condition (1.25), we take
$$x = 1/\sqrt{3} \quad y = -1/\sqrt{3} \quad z = 1/\sqrt{3}.$$
These are the direction cosines of the principal axes about which the moment of inertia is A^*.

Similarly, with $\lambda = \tfrac{11}{12}Ma^2$ and $\tfrac{1}{4}Ma^2$, the corresponding solutions are
$$x = 1/\sqrt{2} \quad y = 0, \quad , \quad z = -1/\sqrt{2},$$
$$x = 1/\sqrt{6}, \quad y = 2/\sqrt{6}, \quad z = 1/\sqrt{6}.$$
These are the direction cosines of the principal axes about which the moments of inertia are B^* and C^* respectively.

The transformation matrix $\mathbf{l}$ relating coordinates with respect to the axes parallel to $Oxyz$ to coordinates with respect to the principal axes is
$$\mathbf{I} = \frac{1}{\sqrt{6}}\begin{bmatrix} \sqrt{2} & -\sqrt{2} & \sqrt{2} \\ \sqrt{3} & 0 & -\sqrt{3} \\ 1 & 2 & 1 \end{bmatrix}.$$

□

9

B

Problem 1.9 Calculate the principal axes and moments of inertia at a corner of a uniform cube.

Solution. Taking axes through the centre of the cube perpendicular to the faces, the moments and products of inertia in this frame are found to be

$$A_G = B_G = C_G = \tfrac{1}{6}Ma^2, \quad F_G = G_G = H_G = 0,$$

where M is the mass of the cube and a is the length of an edge.

Transforming to parallel axes $Oxyz$ through a corner, we now find that

$$A = B = C = \tfrac{2}{3}Ma^2, \quad F = G = H = \tfrac{1}{4}Ma^2.$$

The characteristic equation (1.23) is

$$\begin{vmatrix} \tfrac{2}{3}-\mu & -\tfrac{1}{4} & -\tfrac{1}{4} \\ -\tfrac{1}{4} & \tfrac{2}{3}-\mu & -\tfrac{1}{4} \\ -\tfrac{1}{4} & -\tfrac{1}{4} & \tfrac{2}{3}-\mu \end{vmatrix} = 0,$$

with $\mu = \lambda/Ma^2$. The characteristic roots are $\mu = \tfrac{11}{12}, \tfrac{11}{12}, \tfrac{1}{6}$ and the principal moments of inertia are accordingly

$$A^* = \tfrac{11}{12}Ma^2, \quad B^* = \tfrac{11}{12}Ma^2, \quad C^* = \tfrac{1}{6}Ma^2.$$

Taking $\lambda = \tfrac{11}{12}Ma^2$, the equations (1.24) all become

$$x+y+z = 0. \tag{1.27}$$

This has an infinity of solutions $x:y:z$. Only the third principal axis corresponding to the root $\lambda = \tfrac{1}{6}Ma^2$ is unique; this root leads to equations (1.24) having the form

$$-2x+y+z = 0, \quad x-2y+z = 0, \quad x+y-2z = 0.$$

Thus, $x = y = z = 1/\sqrt{3}$ are the direction cosines of this principal axis; it is clearly the diagonal of the cube through the corner O. Any pair of axes perpendicular to this diagonal can be chosen to represent the two remaining principal axes; all such axes have direction cosines satisfying equation (1.27). $\qquad\square$

EXERCISES

1. A solid is bounded by the coordinate planes and that part of the ellipsoid

$$\frac{x^2}{a^2}+\frac{y^2}{b^2}+\frac{z^2}{c^2} = 1$$

which lies in the first octant. If the density varies as the square of the distance from the origin, show that the z-coordinate of the centre of mass is given by

$$\bar{z} = \frac{5c}{16}\cdot\frac{a^2+b^2+2c^2}{a^2+b^2+c^2}.$$

(Hint: Transform by $x = ar\sin\theta\cos\phi$, $y = br\sin\theta\sin\phi$, $z = cr\cos\theta$.)

2. A circle of radius a is rotated about an axis l in its plane, distant $r(> a)$ from its centre, to generate a torus. If the torus is constructed from uniform material, show that its radius of gyration about l is k, where

$$k^2 = \tfrac{3}{4}a^2 + r^2.$$

3. A lamina lies in the xy-plane and has moments of inertia A, B and product of inertia H with respect to these axes. Axes $Ox'y'$ are obtained from Oxy by rotation in the plane through an angle θ. Show that the lamina's product of inertia with respect to the rotated axes is given by

$$H' = \tfrac{1}{2}(A - B)\sin 2\theta + H\cos 2\theta.$$

Hence find θ if Ox', Oy' are principal axes.

If the lamina is a uniform triangular plate with sides a, b enclosing an angle C, show that a principal axis through the vertex C makes angles θ, ϕ with the two sides such that

$$\tan(\phi - \theta) = \frac{(a^2 - b^2)\sin C}{(a^2 + b^2)\cos C + ab}.$$

4. A uniform solid tetrahedron of mass M is bounded by the coordinate planes and the plane

$$\frac{x}{a} + \frac{y}{b} + \frac{z}{c} = 1.$$

Calculate its moments and products of inertia with respect to the frame $Oxyz$ and hence show that its moment of inertia about an axis through O perpendicular to the opposite face is

$$\tfrac{1}{10}M\left[a^2 + b^2 + c^2 - \frac{6a^2b^2c^2}{b^2c^2 + c^2a^2 + a^2b^2}\right].$$

5. A uniform hemisphere has mass M and radius a. Rectangular axes $Oxyz$ are constructed through a point O on the circumference of its plane face, Oy passing through the centre of this face and Oz being tangential to the circular edge. Calculate the moments and products of inertia of the hemisphere in this frame and hence show that Oz is a principal axis at O and that one of the other principal axes makes an acute angle $\tan^{-1}3$ with Ox and the other makes the same angle with Oy. Show that the principal moments of inertia are $\tfrac{11}{40}Ma^2$, $\tfrac{7}{5}Ma^2$, $\tfrac{61}{40}Ma^2$.

Chapter 2

Plane Motions

2.1 Systems with Few Particles If a dynamical system comprises only a small number of interacting particles, its motion can often be calculated by writing down the equations of motion of the particles separately. These equations of motion follow from Newton's second law, which is only valid for accelerations calculated relative to an *inertial frame*, i.e. a set of axes whose origin is moving uniformly relative to the centre of our galaxy and which is non-rotating relative to the galaxy. Unless otherwise stated, all dynamical quantities (e.g. acceleration, linear momentum, etc.) which arise in the problems which follow, will be supposed calculated within an inertial frame. However, a frame fixed in the earth will often be treated as if it were inertial, with the consequence that the results calculated in such cases will be approximate only.

Problem 2.1 A particle P of mass m lies on a smooth horizontal table and is attached to a long, inextensible string which passes through a smooth hole in the table. The other end of the string carries a particle Q of mass km hanging freely. The particle P is projected from rest along the table at right angles to the string with speed $\sqrt{(8ag)}$, when it is a distance a from the hole. Prove that the particle Q will begin to rise if $k < 8$, and that, if the greatest distance of P from the hole is $2a$, then $k = 3$.

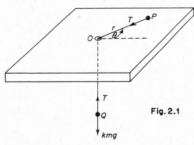

Fig. 2.1

Solution. Let (r, θ) be the polar coordinates of P at time t with respect to the hole as pole (Fig. 2.1). Then, if T is the tension in the string, resolving the equation of motion for P in the radial and transverse directions, we obtain

$$m(\ddot{r} - r\dot{\theta}^2) = -T, \qquad (2.1)$$

$$\frac{m}{r}\frac{d}{dt}(r^2\dot{\theta}) = 0. \qquad (2.2)$$

Since the string is inextensible, the acceleration of Q is $\ddot{r}$ vertically upwards and its equation of motion is

$$km\ddot{r} = T - kmg. \qquad (2.3)$$

Integrating equation (2.2) and employing the initial conditions, we get

$$r^2\dot{\theta} = \text{constant} = a\sqrt{(8ag)}. \qquad (2.4)$$

Eliminating T and $\dot{\theta}$ from equation (2.1) by using equations (2.3), (2.4), the following equation is obtained:

$$(k+1)\ddot{r} = (8a^3g/r^3) - kg. \qquad (2.5)$$

Initially $r = a$ and it follows that at this instant $\ddot{r} = (8-k)g/(k+1)$. Thus $\ddot{r} > 0$ initially, and Q commences to rise provided $k < 8$.

Since $\ddot{r} = d(\tfrac{1}{2}\dot{r}^2)/dr$, equation (2.5) integrates immediately to yield

$$\tfrac{1}{2}(k+1)\dot{r}^2 = (k+4)ag - kgr - (4a^3g/r^2),$$

the constant of integration following from the initial conditions $r = a$, $\dot{r} = 0$. If P is at its maximum distance from the hole when $r = 2a$, $\dot{r}$ must vanish for this value of r also. Thus $k = 3$. ☐

Problem 2.2 Let m_1, m_2 be the masses of a planet and its satellite (or the masses of the components of a binary star system), attracting one another according to the Newtonian law of gravitation. Discuss their motions.

Solution. If $\mathbf{r}_1$, $\mathbf{r}_2$, are the position vectors of the centres of these two bodies with respect to the origin O of an inertial frame, the forces $\mathbf{F}_1$, $\mathbf{F}_2$ exerted upon them are given by

$$\mathbf{F}_1 = \frac{Gm_1 m_2 (\mathbf{r}_2 - \mathbf{r}_1)}{|\mathbf{r}_1 - \mathbf{r}_2|^3}, \qquad \mathbf{F}_2 = \frac{Gm_1 m_2 (\mathbf{r}_1 - \mathbf{r}_2)}{|\mathbf{r}_2 - \mathbf{r}_2|^3},$$

where G is the gravitational constant (Fig. 2.2). Evidently $\mathbf{F}_1 = -\mathbf{F}_2$ by Newton's third law. The equations of motion of the bodies are therefore

$$m_1 \ddot{\mathbf{r}}_1 = \frac{Gm_1 m_2 (\mathbf{r}_2 - \mathbf{r}_1)}{|\mathbf{r}_1 - \mathbf{r}_2|^3}, \qquad m_2 \ddot{\mathbf{r}}_2 = \frac{Gm_1 m_2 (\mathbf{r}_1 - \mathbf{r}_2)}{|\mathbf{r}_1 - \mathbf{r}_2|^3}. \qquad (2.6)$$

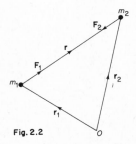

Fig. 2.2

13

Adding these equations, we find that
$$m_1 \ddot{\mathbf{r}}_1 + m_2 \ddot{\mathbf{r}}_2 = 0$$
and hence, if $\bar{\mathbf{r}}$ is the position vector of the centre of mass of the bodies, it follows that
$$(m_1 + m_2)\ddot{\bar{\mathbf{r}}} = 0.$$
Thus, the centre of mass has zero acceleration and moves with uniform velocity in any inertial frame.

Let $\mathbf{r} = \mathbf{r}_2 - \mathbf{r}_1$ be the position vector of the satellite in a frame S whose axes are parallel to the original frame and whose origin is at the centre of the planet. It then follows from equations (2.6) that
$$\ddot{\mathbf{r}} = \ddot{\mathbf{r}}_2 - \ddot{\mathbf{r}}_1 = -G(m_1 + m_2)\mathbf{r}/r^3, \tag{2.7}$$
where $r = |\mathbf{r}|$. This is exactly the equation of satellite motion which would be derived if S were treated as inertial (i.e. the planet were taken to be unaccelerated) and the mass of the planet were taken to be $(m_1 + m_2)$; this we shall term the corrected mass of the planet.

Taking the vector product of both sides of equation (2.7) with $\mathbf{r}$, it follows that
$$d(\mathbf{r} \times \dot{\mathbf{r}})/dt = \mathbf{r} \times \ddot{\mathbf{r}} = 0.$$
Thus
$$\mathbf{r} \times \dot{\mathbf{r}} = \mathbf{h}, \tag{2.8}$$
where $\mathbf{h}$ is a constant vector, and this implies that $\mathbf{r}$ is always perpendicular to $\mathbf{h}$, i.e. that the satellite's trajectory lies in a fixed plane through the planet.

Taking polar coordinates (r, θ) in the plane of the satellite's motion, the planet lying at the pole, the radial and transverse components of equation (2.7) are $(\mu = G(m_1 + m_2))$
$$\ddot{r} - r\dot{\theta}^2 = -\frac{\mu}{r^2}, \quad \frac{1}{r}\frac{d}{dt}(r^2\dot{\theta}) = 0. \tag{2.9}$$
Thus
$$r^2\dot{\theta} = h \quad \text{(a constant)}, \tag{2.10}$$
which is a scalar form of equation (2.8). h is the moment of the satellite's velocity about the pole. Putting $u = 1/r$, then $\dot{\theta} = hu^2$ and hence
$$\dot{r} = -\frac{1}{u^2}\frac{du}{dt} = -\frac{1}{u^2}\frac{du}{d\theta}\dot{\theta} = -h\frac{du}{d\theta},$$
$$\ddot{r} = -\frac{d}{d\theta}\left(h\frac{du}{d\theta}\right)\frac{d\theta}{dt} = -h^2u^2\frac{d^2u}{d\theta^2}.$$
The first of equations (2.9) can now be written in the form
$$\frac{d^2u}{d\theta^2} + u = \frac{\mu}{h^2}.$$

14

This is a linear equation with constant coefficients whose general solution is

$$u = A\cos(\theta - \alpha) + (\mu/h^2),$$

A, α being constants of integration. By appropriate choice of the line $\theta = 0$, α can always be reduced to zero. Thus this equation is equivalent to

$$l/r = 1 + e\cos\theta, \tag{2.11}$$

where $l = h^2/\mu$ and $e = Ah^2/\mu$. Equation (2.11) is the polar equation of a conic of eccentricity e and semilatus rectum l with a focus at the pole. It has accordingly been shown that the orbit of the satellite as observed from the planet is an ellipse (e must be less than one, for otherwise the satellite recedes to an infinite distance along a parabolic or hyperbolic path), with a focus at the planet's centre.

Integrating equation (2.10) over a time interval T during which the satellite completes one circuit of its orbit from $\theta = 0$ to $\theta = 2\pi$ we obtain

$$2A = \int_0^{2\pi} r^2 \, d\theta = hT,$$

where A is the area of the ellipse. But, $A = \pi ab$, where a, b, are the semi-axes of the ellipse and $h = (\mu l)^{\frac{1}{2}} = \mu^{\frac{1}{2}}b/a^{\frac{1}{2}}$. It follows that

$$T = 2\pi a^{\frac{3}{2}}/\mu^{\frac{1}{2}}.$$

This equation permits a calculation of μ from observed values of the periodic time T and the semi-major axis a. It also leads to Kepler's third law, $T^2 \propto a^3$. ▢

2.2 Conservative Systems. Energy Equation The *principle of work* for a particle states that the net work done by the forces acting upon it is equal to the increase in its kinetic energy. This principle can be extended to any system of particles, yielding the result that the net work done by the forces of the system (both internal and external) is equal to the increment in the overall kinetic energy of the system.

In certain cases, the work done by the internal and external forces as the system moves from one configuration to another, depends only upon the end configurations. In these circumstances, the *potential energy* of the system in a given configuration is defined as the work which would be done by the forces of the system if it were to move from the given configuration to some standard configuration (which can be chosen arbitrarily). For such a system, the principle of work can be expressed as the *principle of conservation of energy*,

$$\text{kinetic energy} + \text{potential energy} = \text{constant.} \tag{2.12}$$

15

The kinetic energy of a moving rigid body of mass M can be calculated in two parts, (i) the kinetic energy of the overall translatory motion, which is that of a particle of mass M moving with the centre of mass of the body, and (ii) the kinetic energy of the rotary motion relative to axes through the centre of mass parallel to the axes of the inertial frame being employed. If the motion of the rigid body is parallel to a plane π and $\bar{v}$ is the velocity of its centre of mass, ω is its angular velocity about an axis through the centre of mass perpendicular to π, I_G is its moment of inertia about this axis, then its kinetic energy is given by the formula

$$T = \tfrac{1}{2}M\bar{v}^2 + \tfrac{1}{2}I_G\,\omega^2. \tag{2.13}$$

In the special case of a rigid body rotating about a fixed axis, its kinetic energy can be calculated immediately from the formula

$$T = \tfrac{1}{2}I\omega^2, \tag{2.14}$$

where I is the moment of inertia of the body about the axis and ω is its angular velocity.

Problem 2.3 A uniform rod is placed with one end on a smooth vertical wall and the other on a smooth horizontal ground. It is released and slides in a vertical plane. Determine its motion.

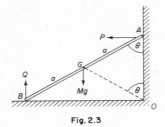

Fig. 2.3

Solution. Let the smooth reactions of the wall and ground on the rod be P and Q as shown in Fig. 2.3. The weight Mg acts through G the centre of the rod and is the only force which does work as the rod moves. The weight is a conservative force and the work done by this force as the rod moves from its position in the figure to a horizontal position on the ground (the standard configuration) is $Mga\cos\theta$ ($2a$ is the length of the rod). This, therefore, is the potential energy of the rod.

A semicircle can be constructed on AB as diameter and will pass through O; hence $OG = GA = GB = a$. Thus, the triangle AGO is isosceles and angle GOA is θ. It follows that G describes a circle, centre O, radius a, with angular velocity $\dot{\theta}$. The angular velocity of the rod about G is also $\dot{\theta}$

16

and its moment of inertia about this point is $\frac{1}{3}Ma^2$. The kinetic energy of the rod is now calculated from equation (2.13) to be

$$\tfrac{1}{2}M(a\dot\theta)^2 + \tfrac{1}{2}\cdot\tfrac{1}{3}Ma^2\cdot\dot\theta^2 = \tfrac{2}{3}Ma^2\dot\theta^2.$$

The energy equation for the rod's motion is therefore

$$\tfrac{2}{3}Ma^2\dot\theta^2 + Mga\cos\theta = \text{constant}.$$

Assuming that the rod's initial inclination to the wall is α, we have $\dot\theta = 0$ when $\theta = \alpha$. This fixes the constant in the last equation, and hence

$$\dot\theta^2 = 3g(\cos\alpha - \cos\theta)/2a. \tag{2.15}$$

This differential equation determines the motion so long as the rod remains in contact with the wall (see Problem 2.5). It can be integrated to yield θ as a function of t, but the process leads to an elliptic integral. $\qquad\square$

Problem 2.4 A uniform circular cylinder of mass M and radius a lies at rest on a perfectly rough horizontal plane. A particle of mass m is fixed to it at a point on its highest generator. If the equilibrium is slightly disturbed, calculate the subsequent motion.

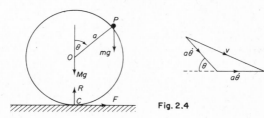

Fig. 2.4

Solution. The external forces acting upon the system (cylinder + particle) are (i) the weight Mg of the cylinder acting at the centre of mass O, (ii) the weight mg of the particle P, (iii) the normal component R of the reaction of the plane and (iv) the frictional component F of this reaction. Of these, only the weight of the particle does work during the motion; F does no work, provided the cylinder does not slip, since the cylinder rotates about C as instantaneous centre until a new particle of the cylinder is brought into contact with the plane and acts as instantaneous centre in its turn; thus the point of application of the frictional force is always stationary. Since weight forces are conservative, the potential energy of the system in the configuration shown in Fig. 2.4 is $mga\cos\theta$. The internal forces, of course, do no work.

When the cylinder has rotated through an angle θ, its angular velocity

17

about O will be $\dot{\theta}$ and the velocity of O will be $a\dot{\theta}$ horizontally. Thus, the kinetic energy of the cylinder is

$$\tfrac{1}{2}M(a\dot{\theta})^2 + \tfrac{1}{2}.\tfrac{1}{2}Ma^2\dot{\theta}^2 = \tfrac{3}{4}Ma^2\dot{\theta}^2,$$

$\tfrac{1}{2}Ma^2$ being its moment of inertia about O. The particle has the velocities $a\dot{\theta}$ perpendicular to OP and $a\dot{\theta}$ horizontally, simultaneously; the triangle of velocities (Fig. 2.4) shows that its resultant velocity is v, where

$$v^2 = 2a^2\dot{\theta}^2(1 + \cos\theta).$$

The energy equation for the system can now be written down:

$$\tfrac{3}{4}Ma^2\dot{\theta}^2 + ma^2\dot{\theta}^2(1 + \cos\theta) + mga\cos\theta = \text{const.}$$

Initially $\theta = 0$, $\dot{\theta} = 0$ and hence const. $= mga$. The equation of motion is accordingly

$$\{\tfrac{3}{4}M + m(1 + \cos\theta)\}a\dot{\theta}^2 = mg(1 - \cos\theta). \tag{2.16}$$

A further integration introduces elliptic integrals.　□

2.3 Motion of the Centre of Mass　The centre of mass of a system of particles moves as if the whole mass of the system were concentrated there and all the external forces were applied there. Thus, if M is the total mass of the system and $\mathbf{F}_E$ are the external forces, then

$$\sum \mathbf{F}_E = M\ddot{\mathbf{r}} = M\, d\bar{\mathbf{v}}/dt. \tag{2.17}$$

The motion of the centre of mass is completely independent of the internal forces between particles of the system.

Problem 2.5　Calculate the forces exerted by the wall and ground on the rod in Problem 2.3.

Solution. The horizontal distance of G from the wall (Fig. 2.3) is $a\sin\theta$. The horizontal component of the acceleration of G is accordingly

$$d^2(a\sin\theta)/dt^2 = a(\cos\theta\,\ddot{\theta} - \sin\theta\,\dot{\theta}^2).$$

Resolving the external forces acting on the rod horizontally, therefore, we get

$$P = Ma(\cos\theta\,\ddot{\theta} - \sin\theta\,\dot{\theta}^2). \tag{2.18}$$

Differentiating equation (2.15) with respect to t, we find that

$$2\dot{\theta}\,\ddot{\theta} = 3g\sin\theta\dot{\theta}/2a.$$

We can now substitute for $\ddot{\theta}$ and $\dot{\theta}^2$ in equation (2.18) and obtain the result

$$P = \tfrac{3}{4}Mg(3\cos\theta - 2\cos\alpha)\sin\theta.$$

This equation shows that the rod loses contact with the wall when $\cos\theta = \tfrac{2}{3}\cos\alpha$, i.e. when the end A has descended to two thirds of its initial height above the ground.

The vertical acceleration of G (upwards) is
$$d^2(a\cos\theta)/dt^2 = -a(\sin\theta\,\ddot\theta + \cos\theta\,\dot\theta^2)$$
$$= -\tfrac{3}{4}g(1 + 2\cos\alpha\cos\theta - 3\cos^2\theta).$$

Resolving the external forces vertically, this leads to the result
$$Q = \tfrac{1}{4}Mg(1 - 6\cos\alpha\cos\theta + 9\cos^2\theta).$$

It is easy to verify that Q is always positive and that the instant the rod leaves the wall, $Q = \tfrac{1}{4}Mg$. ☐

2.4 Conservation of Momentum

In the case where the vector sum of the external forces acting upon a system has zero component in a certain direction, if this direction is taken to be that of the x-axis, equation (2.17) yields
$$M\ddot{\bar{x}} = 0. \tag{2.19}$$

Integrating, it follows that
$$M\dot{\bar{x}} = \text{constant.} \tag{2.20}$$

Equation (1.2) implies that this is equivalent to
$$p_x = M\dot{\bar{x}} = \sum m_i\dot{x}_i = \text{constant,} \tag{2.21}$$

where p_x is the x-component of the net linear momentum $\mathbf{p}$ of the system. This equation is referred to as the *principle of conservation of linear momentum*.

Problem 2.6 Two equal uniform rods AB, BC, each of mass m, are smoothly jointed together at B and lie along a straight line ABC on a smooth horizontal plane. A particle of mass $4m$ is connected to B by a light string which is just taut and lies along a straight line perpendicular to ABC also lying in the plane. The pivot B is held stationary and the rods are given angular velocities ω in opposite senses, so that their ends A, C commence to move away from the particle $4m$. The pivot is released and the system moves on the plane. Calculate the tension in the string when the rods have rotated through an angle θ.

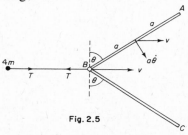

Fig. 2.5

19

Solution. The motion is evidently symmetrical about the line of the string. In the configuration shown in Fig. 2.5, let v be the velocity of the pivot B and, hence, of the mass $4m$. Then the centre of mass of each rod has velocity $a\dot\theta$ relative to B; the velocity of the centre of mass relative to the plane is accordingly found by summing this velocity vectorially with the velocity of B as indicated.

Since no external forces act on the system in its plane of motion, its linear momentum is conserved. The component of the linear momentum of a rod parallel to the string is found from equation (2.21) to be $m(v + a\dot\theta\cos\theta)$ (i.e. mass of rod times the component of the velocity of the centre of mass parallel to the string). We can now write down an equation expressing conservation of momentum for the system in this direction:

$$4mv + 2m(v + a\dot\theta\cos\theta) = \text{constant.}$$

Since, initially, $v = 0$, $\theta = 0$, $\dot\theta = \omega$, this gives

$$v = \tfrac{1}{3}a(\omega - \dot\theta\cos\theta). \tag{2.22}$$

The kinetic energy of a particle of mass m moving with the centre of mass of a rod is

$$\tfrac{1}{2}m(v^2 + a^2\dot\theta^2 + 2a\dot\theta v\cos\theta).$$

The kinetic energy of the rotational motion of a rod relative to its centre of mass is $\tfrac{1}{6}ma^2\dot\theta^2$. We can now construct the energy equation for the system in the form

$$\tfrac{1}{2} \cdot 4mv^2 + m(v^2 + a^2\dot\theta^2 + 2a\dot\theta v\cos\theta) + \tfrac{1}{3}ma^2\dot\theta^2 = \text{const.}$$

Utilizing the initial conditions, this gives

$$3v^2 + \tfrac{4}{3}a^2\dot\theta^2 + 2a\dot\theta v\cos\theta = \tfrac{4}{3}a^2\omega^2. \tag{2.23}$$

Equations (2.22), (2.23), completely fix the behaviour of the system. Eliminating v between them, we find that

$$\dot\theta^2 = \omega^2/(1 + \tfrac{1}{3}\sin^2\theta). \tag{2.24}$$

Then, substituting in equation (2.22), an equation for v emerges,

$$v = \tfrac{1}{3}a\omega[1 - \cos\theta/(1 + \tfrac{1}{3}\sin^2\theta)^{\frac{1}{2}}]. \tag{2.25}$$

Differentiating equation (2.25), we get

$$\dot v = 4a\omega^2\sin\theta/(3 + \sin^2\theta)^2.$$

But if T is the tension in the string, the equation of motion of the particle $4m$ is $T = = 4m\dot v$. Thus, a formula for T in terms of θ now follows immediately. $\qquad\square$

Problem 2.7 A ring of mass m is attached to one end of a uniform rod

20

of length $2a$ and mass $2m$ and is then threaded on to a smooth rigid horizontal wire. The rod is set swinging in the vertical plane containing the wire. Discuss the motion.

Solution. The external forces acting upon the system (rod + ring) are the two weight forces mg, $2mg$ and the smooth reaction R of the wire on the ring. Since all these forces are vertical, the horizontal momentum of the

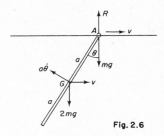

Fig. 2.6

system will be conserved. Taking the velocity of the ring to be v and the angular velocity of the rod to be $\dot{\theta}$ (Fig. 2.6), the equation expressing this principle is

$$mv + 2m(v - a\dot{\theta}\cos\theta) = \text{constant}.$$

We shall suppose that, initially, $\theta = \alpha$, $\dot{\theta} = 0$ and $v = 0$. Then, this equation gives

$$v = \tfrac{2}{3}a\dot{\theta}\cos\theta. \tag{2.26}$$

Conservation of energy for the system is expressed by the equation

$$\tfrac{1}{2}mv^2 + m(v^2 + a^2\dot{\theta}^2 - 2a\dot{\theta}v\cos\theta) + \tfrac{1}{3}ma^2\dot{\theta}^2 - 2mga\cos\theta = \text{constant}.$$

Making use of the initial conditions, this reduces to

$$9v^2 + 8a^2\dot{\theta}^2 - 12a\dot{\theta}v\cos\theta = 12ga(\cos\theta - \cos\alpha). \tag{2.27}$$

Equations (2.26), (2.27) determine the motion. Eliminating v, we get

$$a\dot{\theta}^2(1 + \sin^2\theta) = 3g(\cos\theta - \cos\alpha). \tag{2.28}$$

Thus, if α and θ are small, the oscillations of the rod are governed by the approximate equation

$$\dot{\theta}^2 = 3g(\alpha^2 - \theta^2)/2a, \tag{2.29}$$

all but second order terms having been neglected. Equation (2.29) indicates that small oscillations are synchronous with similar oscillations of a simple pendulum of length $2a/3$.

To calculate R, it is necessary to resolve equation (2.17) for the system vertically. In this case, it is convenient to calculate the right-hand member

21

in two parts, by application of equation (1.4); thus, differentiating this equation twice, we obtain

$$M\ddot{\mathbf{r}} = M_1\ddot{\mathbf{r}}_1 + M_2\ddot{\mathbf{r}}_2, \tag{2.30}$$

where M_1, M_2 denote the masses of the ring and rod respectively and $\ddot{\mathbf{r}}_1$, $\ddot{\mathbf{r}}_2$ are the respective accelerations of the ring's and rod's centres of mass. Since the ring has no vertical acceleration, the equation for R takes the form

$$R - 3mg = 2m(a\dot{\theta}^2\cos\theta + a\ddot{\theta}\sin\theta). \tag{2.31}$$

By differentiating equation (2.28), $\ddot{\theta}$ can be calculated thus

$$a\ddot{\theta} = 3g\sin\theta(2\cos\theta\cos\alpha + \sin^2\theta - 3)/2(1 + \sin^2\theta)^2. \tag{2.32}$$

Then, substituting from equations (2.28), (2.32) into equation (2.31), it is found that

$$R = 3mg(\cos^2\theta - 2\cos\alpha\cos\theta + 2)/(1 + \sin^2\theta)^2. \tag{2.33} \quad \square$$

2.5 Equation of Angular Momentum   The sum of the moments of the linear momenta of a system of particles about a point C, is called the *angular momentum* $\mathbf{h}$ of the system about C. If $\mathbf{r}_i$ is the position vector of the particle m_i relative to C and $\mathbf{v}_i$ is the velocity of this particle in the inertial frame F being used, then

$$\mathbf{h} = \sum_i \mathbf{r}_i \times m_i\mathbf{v}_i. \tag{2.34}$$

If C is the origin O of F, then $\dot{\mathbf{r}}_i = \mathbf{v}_i$ and it follows that

$$\frac{d\mathbf{h}}{dt} = \sum_i \dot{\mathbf{r}}_i \times m_i\mathbf{v}_i + \sum_i \mathbf{r}_i \times m_i\dot{\mathbf{v}}_i$$
$$= \sum_i \mathbf{r}_i \times (\mathbf{F}_{iE} + \mathbf{F}_{iI}) = \sum_i \mathbf{r}_i \times \mathbf{F}_{iE}, \tag{2.35}$$

where $\mathbf{F}_{iE}$, $\mathbf{F}_{iI}$ are the respective external and internal forces acting on m_i. Thus, the rate of change of the angular momentum about O is equal to the sum of the moments of the external forces about O.

Like the energy, $\mathbf{h}$ can always be expressed as the sum of two parts, (i) the angular momentum about O of a particle moving with the centre of mass and of mass equal to that of the whole system and (ii) the angular momentum about the centre of mass of the motion relative to axes through the centre of mass parallel to the axes of F. Thus, we write

$$\mathbf{h} = \mathbf{h}_G + \bar{\mathbf{r}} \times M\bar{\mathbf{v}}, \tag{2.36}$$

where $\bar{\mathbf{r}}$ is the position vector of the centre of mass relative to O, $\bar{\mathbf{v}}$ is the velocity of the centre of mass relative to F and $\mathbf{h}_G$ is the part (ii). If the system is a rigid body, its motion relative to the centre of mass is a rotation with some angular velocity ω.

In the special case of plane motion, $\mathbf{h}$ is always normal to the plane of motion π and only its magnitude need be considered. For a rigid body in plane motion, the angular velocity of its motion relative to the centre of mass is also normal to π and

$$h_G = I_G\omega, \tag{2.37}$$

where I_G is the moment of inertia of the body about the axis normal to π through G.

Differentiating equation (2.36) and noting that $\dot{\mathbf{r}} = \bar{\mathbf{v}}$, it is found that $\dot{\mathbf{h}} = \dot{\mathbf{h}}_G + \bar{\mathbf{r}} \times M\dot{\bar{\mathbf{v}}} = \dot{\mathbf{h}}_G + \sum \bar{\mathbf{r}} \times \mathbf{F}_{iE}$. It now follows from equation (2.35) that

$$\dot{\mathbf{h}}_G = \sum (\mathbf{r}_i - \bar{\mathbf{r}}) \times \mathbf{F}_{iE} = \mathbf{M}_G, \tag{2.38}$$

where $\mathbf{M}_G$ is the sum of the moments of the external forces about G. For the special case of plane motion of a rigid body, equation (2.37) shows that equation (2.38) takes the form

$$I_G\dot{\omega} = M_G. \tag{2.39}$$

Problem 2.8 A uniform cylinder of mass m and radius a rolls inside a fixed cylinder of radius b with its axis horizontal; the cylinders are in contact along a common generator. Discuss the motion.

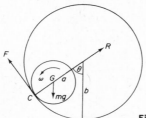

Fig. 2.7

Solution. At time t, let the plane containing the generator of contact and the axis of the moving cylinder be inclined at an angle θ to the vertical (Fig. 2.7). Then, if G is the centre of mass of the moving cylinder, its velocity is $(b-a)\dot{\theta}$ perpendicular to CG. If ω is the angular velocity of this cylinder, the velocity of a point of contact C relative to G is $a\omega$ perpendicular to CG. Thus the velocity of C relative to a fixed observer is $(b-a)\dot{\theta} - a\omega$. This must vanish and hence

$$a\omega = (b-a)\dot{\theta}.$$

Let (F, R) be the frictional and normal components of the reaction at C. Taking moments about G and using equation (2.39), we obtain

$$-Fa = \tfrac{1}{2}ma^2\dot{\omega}.$$

Hence
$$F = -\tfrac{1}{2}m(b-a)\ddot{\theta}.$$

23

The acceleration of G has a component $(b-a)\dot\theta^2$ along CG and a component $(b-a)\ddot\theta$ perpendicular to CG. It follows that

$$R - mg\cos\theta = m(b-a)\dot\theta^2, \qquad (2.40)$$
$$F - mg\sin\theta = m(b-a)\ddot\theta.$$

Hence
$$F = \tfrac{1}{3}mg\sin\theta,$$
$$\ddot\theta = -2g\sin\theta/3(b-a).$$

The last equation shows that the cylinder can oscillate about its lowest position in perfect synchrony with a simple pendulum of length $3(b-a)/2$.

Since $\ddot\theta = d(\tfrac{1}{2}\dot\theta^2)/d\theta$, we can integrate the last equation to give

$$\dot\theta^2 = \Omega^2 - 4g(1-\cos\theta)/3(b-a),$$

where we have supposed $\dot\theta = \Omega$, $\theta = 0$, initially. It now follows from equation (2.40) that

$$R = \tfrac{1}{3}mg(7\cos\theta - 4) + m(b-a)\Omega^2.$$

If the moving cylinder is to make a complete circuit of the fixed cylinder, both $\dot\theta^2$ and R must be positive for $\theta = \pi$. The last two equations indicate that the condition for this is $\Omega^2 > 11g/3(b-a)$. $\qquad\square$

Problem 2.9 A uniform circular cylinder is placed on a horizontal plane and an identical cylinder is balanced on top of it so that the axes of the cylinders are parallel. The equilibrium is slightly disturbed. Determine the motion if there is no slipping.

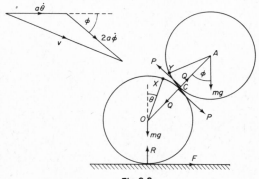

Fig. 2.8

Solution. At time t, let ϕ be the angle made by the plane containing the axes of the cylinders with the vertical and let θ be the angle through which the lower cylinder has rotated. Suppose X, Y are points, one on each cylinder, which were in contact in the equilibrium configuration. Then,

24

if C is the point of contact at time t,
$$\text{angle } YAC = \text{angle } XOC = \phi - \theta.$$

It follows that the angle made by AY with the downward vertical is $2\phi - \theta$ and that the angular velocity of the upper cylinder is $2\dot{\phi} - \dot{\theta}$.

If a is the radius of either cylinder, relative to O, A moves in a circle of radius $2a$ with angular velocity $\dot{\phi}$. O's velocity is $a\dot{\theta}$ horizontally. The velocity v of A in a fixed frame is obtained by compounding these two velocities as indicated in Fig. 2.8; thus
$$v^2 = a^2(\dot{\theta}^2 + 4\dot{\phi}^2 + 4\dot{\theta}\dot{\phi}\cos\phi).$$

The energy equation of the system can now be constructed thus:
$$\tfrac{1}{2}ma^2\dot{\theta}^2 + \tfrac{1}{4}ma^2\dot{\theta}^2 + \tfrac{1}{2}ma^2(\dot{\theta}^2 + 4\dot{\phi}^2 + 4\dot{\theta}\dot{\phi}\cos\phi)$$
$$+ \tfrac{1}{4}ma^2(2\dot{\phi} - \dot{\theta})^2 + 2mga\cos\phi = \text{constant}.$$

Since, initially, $\phi = 0$, $\theta = \dot{\phi} = 0$, this equation reduces to
$$3\dot{\theta}^2 + 6\dot{\phi}^2 + 2\dot{\theta}\dot{\phi}(2\cos\phi - 1) = 4g(1 - \cos\phi)/a. \tag{2.41}$$

Let P be the frictional component of the interaction between the cylinders at C. Taking moments about A for the upper cylinder, we obtain the angular momentum equation
$$Pa = \tfrac{1}{2}ma^2(2\ddot{\phi} - \ddot{\theta}). \tag{2.42}$$

If F is the frictional component of the reaction of the plane upon the lower cylinder, by taking moments about O, an angular momentum equation for the lower cylinder is found:
$$Pa - Fa = \tfrac{1}{2}ma^2\ddot{\theta}. \tag{2.43}$$

The centre of mass of the two cylinders is at C and, relative to O, this point moves on a circle of radius a and centre O. It follows that C has accelerations (i) $a\dot{\phi}^2$ along CO, (ii) $a\ddot{\phi}$ perpendicular to OC and (iii) $a\ddot{\theta}$ horizontally. Resolving the external forces acting upon the whole system horizontally and employing the linear momentum equation (2.17), we now get
$$F = 2m(a\ddot{\theta} + a\ddot{\phi}\cos\phi - a\dot{\phi}^2\sin\phi). \tag{2.44}$$

Eliminating P and F between equations (2.42)–(2.44), the following equation is derived,
$$3\ddot{\theta} = \ddot{\phi}(1 - 2\cos\phi) + 2\dot{\phi}^2\sin\phi = \frac{d}{dt}[\dot{\phi}(1 - 2\cos\phi)]. \tag{2.45}$$

Integration yields the equation
$$3\dot{\theta} = \dot{\phi}(1 - 2\cos\phi). \tag{2.46}$$

25

C

So long as the cylinders remain in contact, the system's behaviour is governed by the equations (2.41), (2.46). By eliminating $\dot\theta$ between these equations, a first order equation for ϕ is obtained,

$$\dot\phi^2(17+4\cos\phi-4\cos^2\phi) = 12g(1-\cos\phi)/a.$$

By resolving horizontally for the external forces acting upon the lower cylinder, an equation for the normal reaction Q between the cylinders at C can also be found and, hence, the value of ϕ when the cylinders separate. ◻

2.6 Conservation of Angular Momentum   In the event that the external forces acting upon a dynamical system have zero net moment about a fixed centre C, equation (2.35) yields the integral

$$\mathbf{h} = \text{constant.} \tag{2.47}$$

This is the *principle of conservation of angular momentum.*

If the external forces have no net moment about a certain axis, e.g. the z-axis, the component of angular momentum about this axis is conserved:

$$h_z = \text{constant.} \tag{2.48}$$

Problem 2.10   A light smooth tube can rotate freely in a horizontal plane about its mid-point O which is fixed. A uniform rod of mass m and length $2a$, is placed inside the tube so that its mid-point M is a distance a from O. The system is given an initial angular velocity ω about the vertical through O. Discuss the motion.

Solution. At time t, let θ be the angle through which the tube has rotated and r the distance of M from O. Since the external forces acting on the system, tube + rod, have no moment about a vertical axis through O, the angular momentum of the system about this axis is conserved. The angular momentum of the rod about O can be calculated from equation (2.36) in the form $\frac{1}{3}ma^2\dot\theta + mr^2\dot\theta$.

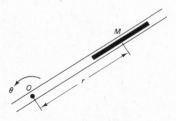

Fig. 2.9

The angular momentum of the tube is negligible, so that the equation expressing conservation of angular momentum is

$$(a^2 + 3r^2)\dot\theta = 4a^2\omega.$$

The principle of conservation of energy leads to the equation

$$\tfrac{1}{2}m(\dot r^2 + r^2\dot\theta^2) + \tfrac{1}{6}ma^2\dot\theta^2 = \tfrac{2}{3}ma^2\omega^2.$$

These two equations determine the motion. Eliminating θ, we find that

$$\dot r = 2a\omega\left(\frac{r^2 - a^2}{3r^2 + a^2}\right)^{\frac{1}{2}},$$

indicating that ultimately, as $r \to \infty$, $\dot r \to 2a\omega/\sqrt{3}$, i.e. the rod moves uniformly relative to the tube. $\qquad\square$

EXERCISES

1. Two particles P and Q, each of mass m, are connected by a light elastic string of modulus λ and natural length a. P is placed close to the edge of a smooth horizontal table and Q is placed on the table so that the string is just taut and perpendicular to an edge. P is pushed gently over the edge. If P, Q move distances x, y respectively in time t, show that

$$x + y = \tfrac{1}{2}gt^2, \quad \ddot z + n^2 z = g,$$

where $z = x - y$ and $n^2 = 2\lambda/ma$. Show that, when the string first becomes slack, Q will still be on the table provided $2\lambda > \pi^2 mg$.

2. A block of mass m is placed with a plane face in contact with a smooth horizontal plane. A circular cylindrical cavity of radius b, with a horizontal axis, exists within the block and a uniform solid circular cylinder of radius a and mass m is free to roll without slipping on the wall of the cavity with its axis parallel to that of the cavity. The system is set in motion so that the solid cylinder executes complete revolutions of the cavity and the block's motion is perpendicular to the cavity axis. If the net linear momentum of the system is zero and contact between the cylinder and the cavity is about to be broken when the cylinder is at its highest point, show that

$$\dot\theta^2(b - a)(2 + \sin^2\theta) = 2g(3 + 2\cos\theta),$$

where θ is the angle made by the plane containing the axes of the solid cylinder and cavity with the downward vertical. Show that the normal and frictional components of the reaction between the cylinder and the cavity wall are given by

$$R = \frac{3mg(1+\cos\theta)(3+\cos\theta)}{(2+\sin^2\theta)^2}, \quad F = \frac{mg\sin\theta(3+3\cos\theta+\cos^2\theta)}{(2+\sin^2\theta)^2}.$$

3.   Two uniform rods AB and BC, each of length $2a$, and having masses m, $2m$ respectively, are smoothly hinged together at B. The rods can move in a vertical plane, the ends A and C sliding on a smooth horizontal plane. If θ is the inclination of AB to the horizontal and $\theta = \alpha$ initially, obtain the equation of motion

$$a\dot\theta^2(12-\sin^2\theta) = 18g(\sin\alpha-\sin\theta).$$

4.   A uniform circular cylinder can turn freely about its axis, which is fixed horizontally. A second identical cylinder is placed with a generator lying along the highest generator of the first cylinder, and the system is given a slight disturbance from rest. If the coefficient of friction between the cylinders is μ, show that slipping begins when the angle θ between the plane of the axes and the upward vertical is such that

$$13\mu\cos\theta-\sin\theta = 8\mu.$$

5.   A bead P of mass m is threaded on a smooth, uniform circular wire of mass m and radius a, free to rotate in a horizontal plane about a smooth vertical axis through a point O of its circumference. With the wire stationary, the bead is projected with velocity u from the point A, where OA is a diameter of the wire. If θ is the angle PCA, where C is the centre of the wire, obtain the equation of motion

$$a^2\dot\theta^2(3-\cos^2\theta) = 2u^2\cos\theta.$$

Chapter 3

Three Dimensional Motions

3.1 Euler's Equations Consider a rigid body which rotates about a fixed point O of itself as pivot. Choosing a rectangular cartesian reference frame $Oxyz$, let $\boldsymbol{\omega} = (\omega_x, \omega_y, \omega_z)$ be the angular velocity of the body about O and let $\mathbf{h} = (h_x, h_y, h_z)$ be its angular momentum about O. Then, if $\mathbf{J}$ is the inertia matrix (equation 1.20) for the body with respect to the frame $Oxyz$, it can be proved that

$$\mathbf{h} = \mathbf{J}\boldsymbol{\omega}, \tag{3.1}$$

or

$$\begin{aligned}
h_x &= A\omega_x - H\omega_y - G\omega_z, \\
h_y &= -H\omega_x + B\omega_y - F\omega_z, \\
h_z &= -G\omega_x - F\omega_y + C\omega_z,
\end{aligned}$$

where the components of $\mathbf{h}$ and $\boldsymbol{\omega}$ are to be arranged as columns. In the special case when the axes of the frame are principal axes, $F = G = H = 0$ and equation (3.1) gives

$$h_x = A\omega_x, \quad h_y = B\omega_y, \quad h_z = C\omega_z, \tag{3.2}$$

where A, B, C are now the principal moments of inertia.

Suppose that the frame $Oxyz$ moves with the body, so that its axes are always principal axes. Then the frame has angular velocity $\boldsymbol{\omega}$ relative to the inertial frame instantaneously coincident with it. Thus, as proved in section 6.4 of *Vector Algebra* (Marder) in this series,

$$\frac{d\mathbf{h}}{dt} = \frac{\partial \mathbf{h}}{\partial t} + \boldsymbol{\omega} \times \mathbf{h}, \tag{3.3}$$

where $d\mathbf{h}/dt$ denotes the rate of change of $\mathbf{h}$ relative to the inertial frame and $\partial\mathbf{h}/\partial t$ denotes the rate of change of $\mathbf{h}$ relative to the rotating frame. Let $\mathbf{M}$ be the moment about O of the external forces applied to the body. Then equations (2.35), (3.3) imply that

$$(\partial\mathbf{h}/\partial t) + \boldsymbol{\omega} \times \mathbf{h} = \mathbf{M}. \tag{3.4}$$

Substituting for $\mathbf{h}$ from equations (3.2), we derive *Euler's equations*,

$$\left.\begin{aligned}
A\dot{\omega}_x + (C-B)\omega_y\omega_z &= M_x, \\
B\dot{\omega}_y + (A-C)\omega_z\omega_x &= M_y, \\
C\dot{\omega}_z + (B-A)\omega_x\omega_y &= M_z.
\end{aligned}\right\} \tag{3.5}$$

Problem 3.1 Discuss the motion of the earth about its centre.

29

Solution. The earth's centre can be treated as a stationary pivot O. Taking Oz through the poles and Ox, Oy through points on the equator, since the earth is very nearly an oblate spheroid, these will be principal axes with $A = B$ and $C > A$. Neglecting tidal forces due to the sun and moon, the motion about O takes place under zero external forces. Thus Euler's equations give

$$A\dot{\omega}_x + (C-A)\omega_y\,\omega_z = 0, \qquad A\dot{\omega}_y - (C-A)\omega_z\,\omega_x = 0, \qquad C\dot{\omega}_z = 0.$$

Thus $\omega_z = n$ (a constant) and

$$\dot{\omega}_x + v\omega_y = \dot{\omega}_y - v\omega_x = 0$$

where $v = n(C-A)/A$. These equations have the general solution

$$\omega_x = a\cos(vt+\alpha), \qquad \omega_y = a\sin(vt+\alpha).$$

Thus

$$h_x = Aa\cos(vt+\alpha), \qquad h_y = Aa\sin(vt+\alpha), \qquad h_z = Cn.$$

These equations show that the angular momentum vector **h** can be represented by OP in Fig. 3.1; in this diagram, $ON = Aa$, $PN = Cn$ and ON

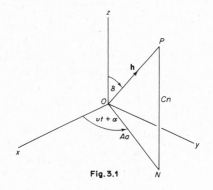

Fig. 3.1

makes an angle $vt+\alpha$ with Ox. It is now clear that OP traces out a cone with Oz as axis and semi-vertical angle δ, where $\tan\delta = Aa/Cn$. But, since the motion takes place under no external forces, equation (2.35) shows that **h** is a constant vector in a non-rotating frame, i.e. OP is a fixed direction in space. Relative to a non-rotating frame, therefore, the earth's axis Oz rotates about a fixed axis OP, thus generating a cone in space with semi-vertical angle δ. This is the phenomenon of *Eulerian nutation*. The rate of rotation is evidently v, so that a complete cycle is completed in time

$$2\pi/v = 2\pi A/n(C-A).$$

As the earth's axis describes its cone, the earth rotates about this axis with angular velocity n. $\qquad\square$

Problem 3.2 Discuss the motion of a body moving freely about a fixed pivot O under no other forces.

Solution. Multiplying Euler's equations (3.5), with $M_x = M_y = M_z = 0$, by ω_x, ω_y, ω_z respectively and adding, we get

$$A\omega_x \dot\omega_x + B\omega_y \dot\omega_y + C\omega_z \dot\omega_z = 0.$$

Upon integration, this yields

$$A\omega_x^2 + B\omega_y^2 + C\omega_z^2 = 2T, \tag{3.6}$$

where T is a constant which will later be identified with the kinetic energy.

Again, multiplying Euler's equations by $A\omega_x$, $B\omega_y$, $C\omega_z$, respectively and adding, we obtain

$$A^2\omega_x \dot\omega_x + B^2\omega_y \dot\omega_y + C^2\omega_z \dot\omega_z = 0.$$

This integrates to give

$$A^2\omega_x^2 + B^2\omega_y^2 + C^2\omega_z^2 = H^2. \tag{3.7}$$

H is the magnitude of the angular momentum vector $\mathbf{h}$ (see equations 3.2).

Equations (3.6), (3.7), can now be solved for ω_x, ω_z in terms of ω_y; then, substituting in the second Euler equation, a first order differential equation for ω_y is obtained. In general, the solution of this equation involves elliptic functions; however, if the initial conditions are chosen appropriately, a solution in elementary terms can arise. For example, if $H^2 = 2BT$, it will be found that

$$\frac{A\omega_x^2}{C-B} = \frac{2T - B\omega_y^2}{C-A} = \frac{C\omega_z^2}{B-A}, \tag{3.8}$$

where we are assuming $A < B < C$. These equations permit us to reduce the second Euler equation to the form

$$\dot\omega_y = \alpha(\lambda^2 - \omega_y^2),$$

where $\qquad \alpha = [(B-A)(C-B)/AC]^{\frac{1}{2}}, \qquad \lambda^2 = 2T/B.$

Choosing the origin of time appropriately, this equation integrates to yield

$$\omega_y = \lambda \tanh \lambda \alpha t.$$

Substituting in equations (3.8), this leads to

$$\omega_x = \left[\frac{B(C-B)}{A(C-A)}\right]^{\frac{1}{2}} \lambda \operatorname{sech} \lambda \alpha t, \qquad \omega_z = \left[\frac{B(B-A)}{C(C-A)}\right]^{\frac{1}{2}} \lambda \operatorname{sech} \lambda \alpha t.$$

Clearly, as $t \to \infty$, $\omega_y \to \lambda$ and $\omega_x \to 0$, $\omega_y \to 0$, i.e. ultimately, the body rotates about the principal axis of inertia corresponding to the intermediate moment of inertia B, with uniform angular velocity λ. □

3.2 Kinetic Energy of a Rigid Body If the body rotates about a fixed pivot O relative to an inertial frame $Oxyz$ with angular velocity ω, its kinetic energy will be given by the equation

$$T = \tfrac{1}{2}I\omega^2, \tag{3.9}$$

where I is the body's moment of inertia about the instantaneous axis of rotation. If this axis has direction cosines (l, m, n), I is given by equation (1.9). Since $\omega_x = l\omega$, $\omega_y = m\omega$, $\omega_z = n\omega$, equation (3.9) can be written in the form

$$T = \tfrac{1}{2}(A\omega_x^2 + B\omega_y^2 + C\omega_z^2 - 2F\omega_y\omega_z - 2G\omega_z\omega_x - 2H\omega_x\omega_y). \tag{3.10}$$

In the special case when the axes Ox, Oy, Oz, are principal axes, this formula reduces to

$$T = \tfrac{1}{2}(A\omega_x^2 + B\omega_y^2 + C\omega_z^2). \tag{3.11}$$

The kinetic energy of a rigid body moving generally can now be calculated by separating this into two parts, (i) the kinetic energy of the translatory motion, and (ii) the kinetic energy of the rotary motion about the centre of mass, as explained in section 2.2.

Problem 3.3 A uniform square plate has one corner freely pivoted to a fixed point O of a smooth horizontal plane. The plate is initially vertical and then falls from rest under gravity, its lower edge sliding on the plane. Calculate the motion.

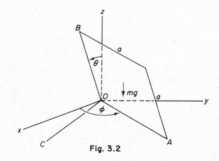

Fig. 3.2

Solution. Let Ox be the initial position on the plane of the edge OA (Fig. 3.2) and let Oy also lie in this plane. Oz is vertically upwards. At time t, let ϕ be the angle made by OA with Ox and let θ be the inclination of the plate to Oz. The plate has a component $\dot{\theta}$ of angular velocity about OA and a component $\dot{\phi}$ about Oz; relative to axes OA, OB, OC, therefore, the components of the angular velocity of the plate are $(\dot{\theta}, \dot{\phi}\cos\theta, -\dot{\phi}\sin\theta)$

32

respectively. Relative to the same axes $OABC$, the moments and products of inertia are

$$A = B = \tfrac{1}{3}ma^2, \qquad C = \tfrac{2}{3}ma^2,$$
$$F = G = 0, \qquad H = \tfrac{1}{4}ma^2,$$

m being the mass of the plate and a the length of an edge. Using equation (3.10), the energy equation for the motion can now be written down:

$$\dot{\theta}^2 + \dot{\phi}^2(1 + \sin^2\theta) - \tfrac{3}{2}\dot{\theta}\dot{\phi}\cos\theta = 3g(1 - \cos\theta)/a. \qquad (3.12)$$

A second equation is needed to determine the motion. Since the weight and the reactions of the pivot and plane have no moment about Oz, equation (2.35) requires that h_z should be conserved. We first calculate the components of $\mathbf{h}$ in the directions of the axes $OABC$ by using equation (3.1) and obtain

$$h_A = ma^2(\tfrac{1}{3}\dot{\theta} - \tfrac{1}{4}\dot{\phi}\cos\theta), \qquad h_B = ma^2(-\tfrac{1}{4}\dot{\theta} + \tfrac{1}{3}\dot{\phi}\cos\theta),$$
$$h_C = -\tfrac{2}{3}ma^2\dot{\phi}\sin\theta.$$

Resolving these along the axes $Oxyz$, leads to the result

$$h_z = ma^2[-\tfrac{1}{4}\dot{\theta}\cos\theta + \tfrac{1}{3}\dot{\phi}(1 + \sin^2\theta)].$$

The equation of conservation of h_z is accordingly

$$\dot{\theta}\cos\theta = \tfrac{4}{3}\dot{\phi}(1 + \sin^2\theta). \qquad (3.13)$$

Equations (3.12), (3.13), determine the motion. Eliminating $\dot{\phi}$ between these equations, it is found that

$$a\dot{\theta}^2(7 + 25\sin^2\theta) = 48g(1 - \cos\theta)(1 + \sin^2\theta).$$

In principle, this can be integrated to give θ as a function of t; then, substituting into equation (3.13), the dependence of ϕ upon t follows. $\qquad\square$

3.3 Motion of Tops

Problem 3.4 One end of the axis of a top is freely pivoted at a fixed point O and the top is set spinning. The moment of inertia of the top about its axis is C and the moment of inertia about any perpendicular axis through O is A. Calculate the equation of angular momentum.

Solution. This problem can be analysed by constructing Euler's equations and the energy equation. We prefer, however, to use it to illustrate the application of direct vector methods, to the solution of dynamical problems.

In Fig. 3.3, $Oxyz$ is a fixed rectangular frame with Oz vertically upwards. The unit vector along the axis of the top is $\mathbf{a} = \mathbf{OA}$. If $\boldsymbol{\omega}$ is the vector angular velocity of the top about O, the velocity of A will be $\boldsymbol{\omega} \times \mathbf{a}$. But this velocity is also $\dot{\mathbf{a}}$ and hence

$$\boldsymbol{\omega} \times \mathbf{a} = \dot{\mathbf{a}}.$$

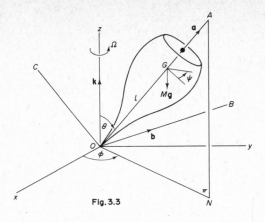

Fig. 3.3

Taking the vector product of both sides of this equation with $\mathbf{a}$ and expanding the triple product, we get

$$\mathbf{a}^2\omega - \mathbf{a}.\omega\mathbf{a} = \mathbf{a} \times \dot{\mathbf{a}}.$$

Since $\mathbf{a}^2 = 1$, this yields

$$\omega = \mathbf{a} \times \dot{\mathbf{a}} + n\mathbf{a},$$

where $n = \mathbf{a}.\omega$ is the component of the top's angular velocity in the direction of the axis.

Resolving along principal axes at O (i.e. the top's axis and any two perpendicular axes), the angular momentum of the top about O is given by

$$\begin{aligned} \mathbf{h} &= (A\omega_1, A\omega_2, Cn) \\ &= A(\omega_1, \omega_2, n) + (C-A)(0, 0, n) \\ &= A\omega + (C-A)n\mathbf{a} \\ &= A\mathbf{a} \times \dot{\mathbf{a}} + Cn\mathbf{a}. \end{aligned}$$

If $\mathbf{g}$ is the vector acceleration due to gravity, the weight of the top is $M\mathbf{g}$; this acts at the centre of mass G on the axis of the top distant l from O. The only other external force acting upon the top is the reaction at the pivot O. Hence, taking moments of these forces about O and employing equation (2.35), we obtain

$$l\mathbf{a} \times M\mathbf{g} = \frac{d}{dt}(A\mathbf{a} \times \dot{\mathbf{a}} + Cn\mathbf{a}) = A\mathbf{a} \times \ddot{\mathbf{a}} + C\dot{n}\mathbf{a} + Cn\dot{\mathbf{a}}.$$

(N.B. $\dot{\mathbf{a}} \times \dot{\mathbf{a}} = \mathbf{0}$.) This is the equation of angular momentum.

Differentiation of the equation $\mathbf{a}^2 = 1$ yields $\mathbf{a}.\dot{\mathbf{a}} = 0$. Thus, if the scalar product of both members of the equation of angular momentum is taken with $\mathbf{a}$, we find that $C\dot{n} = 0$. (N.B. Both the scalar triple products

34

[**aag**], [**aaä**], vanish.) Thus n is constant and the angular momentum equation reduces to

$$A\mathbf{a} \times \ddot{\mathbf{a}} + Cn\dot{\mathbf{a}} = Ml\mathbf{a} \times \mathbf{g}. \tag{3.14}$$

This equation possesses a steady state solution in which the top axis rotates with constant angular velocity Ω about Oz, making a constant angle θ with the vertical. This motion is termed *steady precession*. For such a motion, the point A describes a horizontal circle of radius $\sin\theta$ with angular velocity Ω. Its velocity $\dot{\mathbf{a}}$ is accordingly $\Omega\sin\theta\,\mathbf{b}$, where $\mathbf{b}$ is a unit vector normal to the vertical plane containing the axis of the top ($\mathbf{b} = \mathbf{OB}$ in Fig. 3.3). The acceleration $\ddot{\mathbf{a}}$ of A is $\Omega^2\sin\theta$ along the inwards radius of A's path. $\mathbf{a} \times \ddot{\mathbf{a}}$ can now be calculated to be the vector $-\Omega^2\sin\theta\cos\theta\,\mathbf{b}$ and $\mathbf{a} \times \mathbf{g}$ to be $g\sin\theta\,\mathbf{b}$. Thus, equation (3.14) is satisfied provided θ and Ω are related by

$$-A\Omega^2\sin\theta\cos\theta + Cn\Omega\sin\theta = Mlg\sin\theta,$$

i.e.
$$A\Omega^2\cos\theta - Cn\Omega + Mlg = 0. \tag{3.15}$$

Provided $C^2n^2 \geqslant 4AMgl\cos\theta$, this gives two possible angular velocities of steady precession for a given inclination θ of the axis to the vertical and a given rate of spin n of the top. □

Problem 3.5 Show that, in the general motion of a top, the axis moves between two cones.

Solution. Taking the vector product of both sides of equation (3.14) with $\dot{\mathbf{a}}$, and expanding the triple products, we get

$$A\dot{\mathbf{a}}.\ddot{\mathbf{a}}\mathbf{a} = Ml\dot{\mathbf{a}}.\mathbf{ga}.$$

Thus
$$A\dot{\mathbf{a}}.\ddot{\mathbf{a}} - Ml\dot{\mathbf{a}}.\mathbf{g} = 0,$$

and this equation integrates immediately to give

$$\tfrac{1}{2}A\dot{\mathbf{a}}^2 - Ml\mathbf{a}.\mathbf{g} = E,$$

where E is constant. This is the energy equation. If ϕ is the angle made by the vertical plane π containing the top's axis with the plane Oxz, A will have velocity components (i) $\dot{\theta}$ perpendicular to OA and in the plane π, (ii) $\dot{\phi}\sin\theta$ in the direction of $\mathbf{b}$. Hence

$$\dot{\mathbf{a}}^2 = \dot{\theta}^2 + \dot{\phi}^2\sin^2\theta$$

and the energy equation can be written

$$\tfrac{1}{2}A(\dot{\theta}^2 + \dot{\phi}^2\sin^2\theta) + Mgl\cos\theta = E. \tag{3.16}$$

If $\mathbf{k}$ is the unit vector along Oz, since $\mathbf{k}$ and $\mathbf{g}$ are parallel, taking scalar products of both members of equation (3.14) with $\mathbf{k}$, we get

$$A\mathbf{k}.\mathbf{a} \times \ddot{\mathbf{a}} + Cn\mathbf{k}.\dot{\mathbf{a}} = 0,$$

35

or
$$\frac{d}{dt}(A\mathbf{k}.\mathbf{a}\times\dot{\mathbf{a}}+Cn\mathbf{k}.\mathbf{a}) = 0.$$

Now $\mathbf{k}\times\mathbf{a} = \sin\theta\,\mathbf{b}$ and, hence, $[\mathbf{ka\dot{a}}] = \dot{\phi}\sin^2\theta$. Thus, integrating the last equation, we get

$$A\dot{\phi}\sin^2\theta + Cn\cos\theta = H, \qquad (3.17)$$

where H is constant. Eliminating $\dot{\phi}$ between equations (3.16), (3.17), we obtain

$$A\dot{\theta}^2 = 2E - 2Mgl\cos\theta - \frac{(H-Cn\cos\theta)^2}{A\sin^2\theta}. \qquad (3.18)$$

Putting $\lambda = \cos\theta$, this can be written

$$\dot{\lambda}^2 = (a-b\lambda)(1-\lambda^2)-(c-d\lambda)^2 \equiv f(\lambda),$$

where $a = 2E/A, b = 2Mgl/A, c = H/A$ and $d = Cn/A$.

$f(\lambda)$ is a cubic in λ which must be positive for some value of λ in the interval $(-1,1)$ (otherwise $\dot{\lambda}$ is imaginary for all values of θ). Let this value be λ_0. Then

$$f(-1) < 0, \quad f(\lambda_0) > 0, \quad f(1) < 0, \quad f(+\infty) > 0.$$

It follows that the equation $f(\lambda) = 0$ has three real roots in the intervals $(-1,\lambda_0), (\lambda_0, 1), (1,\infty)$. If λ_1, λ_2 are the roots in the first two intervals, λ must always satisfy the inequalities $\lambda_1 \leqslant \lambda \leqslant \lambda_2$. Since $\dot{\lambda}$ can only vanish for $\lambda = \lambda_1, \lambda = \lambda_2$, the sign of $\dot{\lambda}$ can only change for these values of λ. Thus, λ will oscillate between these extreme values and θ will oscillate between the corresponding values θ_1, θ_2. The behaviour of the top's axis is now clear; it will precess about the vertical Oz with angular velocity $\dot{\phi}$ determined by equation (3.17) and, at the same time, it will oscillate between two cones with common vertex O, common axis Oz and semivertical angles θ_1, θ_2. $\qquad\qquad\square$

3.4 Rolling and Slipping

Problem 3.6 A uniform sphere moves on a rough horizontal plane. Determine its motion (i) when it rolls, (ii) when it slips.

Solution. $Oxyz$ is a frame fixed in the horizontal plane as shown in Fig. 3.4. Let (P, Q, R) be the components of the reaction of the plane on the sphere at its point of contact C and let $(u, v, 0)$ be the components of the velocity of its centre G. Then, if m is the mass of the sphere, the equations of motion of G are

$$P = m\dot{u}, \qquad Q = m\dot{v}, \qquad R-mg = 0. \qquad (3.19)$$

Let $(\omega_1,\omega_2,\omega_3)$ be the components of the angular velocity of the sphere about G. Since the axes parallel to $Oxyz$ through G are always

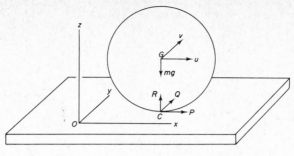

Fig. 3.4

principal axes for the sphere, its angular momentum about G is given by $\mathbf{h}_G = \frac{2}{5}ma^2(\omega_1, \omega_2, \omega_3)$, a being the sphere's radius. Taking moments of the external forces about G and employing equation (2.38), it follows that

$$\tfrac{2}{5}ma^2(\dot\omega_1, \dot\omega_2, \dot\omega_3) = (Qa, -Pa, 0).$$

Thus, ω_3 is constant, and

$$P = -\tfrac{2}{5}ma\dot\omega_2, \qquad Q = \tfrac{2}{5}ma\dot\omega_1. \tag{3.20}$$

If the sphere is rolling upon the plane, C must be stationary. But, the velocity of C relative to G has components $(-a\omega_2, a\omega_1, 0)$. Since G has velocity $(u, v, 0)$, the velocity of C is $(u-a\omega_2, v+a\omega_1, 0)$ and the rolling conditions are

$$u - a\omega_2 = v + a\omega_1 = 0. \tag{3.21}$$

Equations (3.19)–(3.21) determine the rolling motion. Solving, it will be found that $\dot u = \dot v = 0$ and $P = Q = 0$, i.e. the sphere rolls in a straight line with constant speed.

If the sphere is slipping, let θ be the angle made by the velocity of the point C on the sphere with the x-axis. Then

$$(v + a\omega_1)/(u - a\omega_2) = \tan\theta. \tag{3.22}$$

We shall assume that the horizontal (or frictional) component of the reaction at C is in the opposite direction to this velocity and is of magnitude μR, where μ is the coefficient of friction. Hence

$$P = -\mu R \cos\theta, \qquad Q = -\mu R \sin\theta. \tag{3.23}$$

Equations (3.19), (3.20), (3.22), (3.23), determine the motion. Thus

$$\frac{\dot v + a\dot\omega_1}{\dot u - a\dot\omega_2} = \frac{7Q/2m}{7P/2m} = \tan\theta = \frac{v + a\omega_1}{u - a\omega_2}$$

and this implies that

$$\log(v + a\omega_1) = \log(u - a\omega_2) + \text{constant}.$$

37

Hence
$$\frac{v + a\omega_1}{u - a\omega_2} = \tan \theta = \text{constant}, \tag{3.24}$$

i.e. the direction of slip never varies. Equations (3.19), (3.23), now give

$$\dot{u} = -\mu g \cos \theta, \qquad \dot{v} = -\mu g \sin \theta, \tag{3.25}$$

proving that the centre of the sphere moves with uniform acceleration μg in a direction opposite to that of the velocity of slip; this means that G will describe a parabola with its axis aligned with the direction of slip.

Equations (3.20), (3.23), show that

$$\dot{\omega}_1 = -\frac{5\mu g}{2a} \sin \theta, \qquad \dot{\omega}_2 = \frac{5\mu g}{2a} \cos \theta. \tag{3.26}$$

Thus ω_1, ω_2 increase or decrease uniformly. Ultimately, the rolling conditions (3.21) are satisfied and the sphere then begins to roll with constant velocity. ◻

3.5 Rotating Axes
In this section, we shall solve a number of problems on the motion of a rigid body by calculating the components of linear and angular momentum in a rotating frame F whose axes are always principal axes for the body. If $\mathbf{x}$ is a vector whose components with respect to F are known, it is necessary to distinguish between $\partial \mathbf{x}/\partial t$, the rate of change of the vector relative to F and $d\mathbf{x}/dt$, the rate of change of the vector relative to a non-rotating frame instantaneously coincident with F. If $\boldsymbol{\Omega}$ is the angular velocity of F, it is proved in *Vector Algebra* (Marder, section 6.4) that

$$\frac{d\mathbf{x}}{dt} = \frac{\partial \mathbf{x}}{\partial t} + \boldsymbol{\Omega} \times \mathbf{x}. \tag{3.27}$$

Taking $\mathbf{x} = \bar{\mathbf{v}}, \mathbf{h}_G$, this equation enables us to calculate $d\bar{\mathbf{v}}/dt$, $d\mathbf{h}_G/dt$, and use can then be made of equations (2.17), (2.38), to give the equations of linear and angular momentum respectively for the body.

Problem 3.7 A wheel of mass m and radius a rolls on a rough horizontal plane, its centre describing a circle uniformly. Discuss the motion.

Solution. Take axes $G123$ through the centre of the wheel as follows: $G1$ passes through the point of contact C with the plane; $G2$ is horizontal and in the plane of the wheel; $G3$ is perpendicular to the plane of the wheel (Fig. 3.5). These axes are principal axes for the wheel; the associated moments of inertia will be taken to be $(mk^2, mk^2, 2mk^2)$.

Let G move in a circle, centre O and radius c, with angular velocity Ω. Then G has an acceleration $c\Omega^2$ towards O and, if (R, F) are the normal and frictional components of the reaction at C, equation (2.17) gives

$$F = mc\Omega^2, \qquad R = mg. \tag{3.28}$$

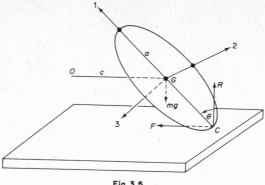

Fig. 3.5

The frame $G123$ rotates about a vertical axis with angular velocity Ω, and hence the components of its angular velocity in this frame are given by $\Omega = (\Omega \cos \theta, 0, -\Omega \sin \theta)$ (θ is the constant inclination of the wheel to the vertical). Let v be the angular velocity of the wheel about its axis for motion relative to the frame $G123$. Then the net angular velocity of the wheel has components $\omega = (\Omega \cos \theta, 0, v - \Omega \sin \theta)$. The angular momentum of the wheel about G is now given by equations (3.2) to be

$$\mathbf{h}_G = mk^2(\Omega \cos \theta, 0, 2n),$$

where $n = v - \Omega \sin \theta$. Hence

$$\frac{d\mathbf{h}_G}{dt} = mk^2 \frac{\partial}{\partial t}(\Omega \cos \theta, 0, 2n) + (\Omega \cos \theta, 0, -\Omega \sin \theta) \times mk^2(\Omega \cos \theta, 0, 2n)$$

$$= mk^2(0, -\Omega^2 \sin \theta \cos \theta - 2n\Omega \cos \theta, 0).$$

Taking moments of the external forces about $G2$, equation (2.38) leads to the result

$$Ra \sin \theta - Fa \cos \theta = mk^2 \Omega \cos \theta (\Omega \sin \theta + 2n). \tag{3.29}$$

In the frame $G123$, C is the point with position vector $(-a, 0, 0)$. Its velocity relative to G is, therefore, $\omega \times \mathbf{a} = (0, -an, 0)$. G has velocity $(0, c\Omega, 0)$ and the net velocity of C is accordingly $(0, c\Omega - an, 0)$. But C is stationary if the wheel does not slip and hence

$$c\Omega = an \tag{3.30}$$

is the rolling condition. Eliminating F, R and n from equation (3.29) by the use of equations (3.28), (3.30), the following relationship between Ω and θ is derived,

$$\Omega^2 = \frac{ga^2 \tan\theta}{k^2(a \sin \theta + 2c) + a^2 c}. \tag{3.31} \quad \square$$

39

Problem 3.8 A gyroscope is mounted in gimbals so that its axis is free to rotate in a horizontal plane. Show that it behaves like a compass.

Solution. $G1EN$ are axes through the mass centre of the flywheel such that $G1$ is vertically upwards and GE, GN lie in a horizontal plane pointing due east and north respectively. Let $G3$ be the axis of the flywheel and let $G2$ be taken horizontally to complete the frame $G123$ (Fig. 3.6). Let λ be

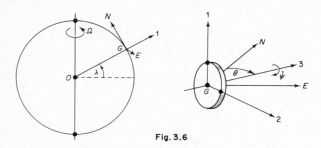

Fig. 3.6

the latitude of G and θ the inclination of $G3$ to GN. The frame $G123$ has angular velocities (i) Ω about an axis parallel to the earth's axis due to the earth's rotation and (ii) $-\dot{\theta}$ about $G1$ due to the motion of the gyroscope's axis. Resolving these angular velocities along the axes of the frame, we obtained the components $(\Omega \sin \lambda - \dot{\theta}, -\Omega \cos \lambda \sin \theta, \Omega \cos \lambda \cos \theta)$ for the net angular velocity of the frame $G123$.

Let ψ be the angle through which the flywheel has rotated with respect to the frame $G123$. Then, the angular velocity of the flywheel has components $(\Omega \sin \lambda - \dot{\theta}, -\Omega \cos \lambda \sin \theta, \dot{\psi} + \Omega \cos \lambda \cos \theta)$ and its angular momentum about G is given by

$$\mathbf{h}_G = [A(\Omega \sin \lambda - \dot{\theta}), -A\Omega \cos \lambda \sin \theta, Cn],$$

where A, A, C are the principal moments of inertia in the frame $G123$ and $n = \dot{\psi} + \Omega \cos \lambda \cos \theta$.

Relative to a non-rotating frame instantaneously coincident with $G123$, the rate of change of $\mathbf{h}_G$ is

$$\frac{d\mathbf{h}_G}{dt} = \frac{\partial}{\partial t}[A(\Omega \sin \lambda - \dot{\theta}), -A\Omega \cos \lambda \sin \theta, Cn]$$

$$+ (\Omega \sin \lambda - \dot{\theta}, -\Omega \cos \lambda \sin \theta, \Omega \cos \lambda \cos \theta)$$
$$\times [A(\Omega \sin \lambda - \dot{\theta}), -A\Omega \cos \lambda \sin \theta, Cn]$$
$$= (-A\ddot{\theta}, -A\Omega \cos \lambda \cos \theta \dot{\theta}, C\dot{n})$$
$$+ [-Cn\Omega \cos \lambda \sin \theta + A\Omega^2 \cos^2 \lambda \sin \theta \cos \theta,$$
$$(A\Omega \cos \lambda \cos \theta - Cn)(\Omega \sin \lambda - \dot{\theta}), 0].$$

Thus the equations of angular momentum are
$$A\ddot\theta + Cn\Omega\cos\lambda\sin\theta - A\Omega^2\cos^2\lambda\sin\theta\cos\theta = 0,$$
$$-2A\Omega\cos\lambda\cos\theta\dot\theta + A\Omega^2\cos\lambda\sin\lambda\cos\theta - Cn(\Omega\sin\lambda - \dot\theta) = M,$$
$$C\dot n = 0,$$

where M is the moment about $G2$ of the forces exerted by the gimbals (these forces have no moments about $G1$ or $G3$).

The first of these equations can be approximated to read

$$A\ddot\theta + Cn\Omega\cos\lambda\sin\theta = 0, \tag{3.32}$$

since Ω^2 will be small by comparison with $n\Omega$. The third equation shows that n is constant.

Equation (3.32) indicates that the axis of the gyroscope will oscillate in a horizontal plane about the north line $\theta = 0$ and that the period of small oscillations is $2\pi\sqrt{(A/Cn\Omega\cos\lambda)}$. These oscillations will be damped by friction and the axis will ultimately come to rest pointing due north. In practice, the frictional couple about $G3$ tending to reduce the spin n is cancelled by the torque of an electric motor. □

Problem 3.9   A sphere of mass m, radius b and radius of gyration k about any diameter, rolls, without slipping, on the outside of a fixed sphere of radius a. Determine the possible motions.

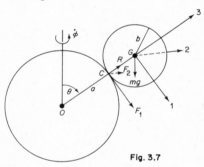

Fig. 3.7

Solution. Take axes $G123$ through the centre of the moving sphere, where $G3$ is along the common radius OG of the two spheres, $G1$ lies in the vertical plane containing this radius and $G2$ is horizontal (Fig. 3.7). θ is the angle between OG and the upward vertical and ϕ is the angle made by the vertical plane $G13$ with a fixed vertical plane. (F_1, F_2, R) are the components of the reaction of the fixed sphere on the moving sphere at the point of contact C. $\omega = (\omega_1, \omega_2, \omega_3)$ is the angular velocity of the moving sphere.

41

The velocity of G is given by
$$\bar{\mathbf{v}} = (a+b)(\dot{\theta}, \dot{\phi}\sin\theta, 0). \tag{3.33}$$
Since C is the point $\mathbf{b} = (0, 0, -b)$ in the frame $G123$, its velocity is
$$\bar{\mathbf{v}} + \boldsymbol{\omega} \times \mathbf{b} = [(a+b)\dot{\theta} - b\omega_2, (a+b)\dot{\phi}\sin\theta + b\omega_1, 0].$$
If there is no slipping, we accordingly have the rolling conditions
$$b\omega_1 = -(a+b)\dot{\phi}\sin\theta, \quad b\omega_2 = (a+b)\dot{\theta}. \tag{3.34}$$
The frame $G123$ has simultaneous angular velocities (i) $\dot{\theta}$ about $G2$ and (ii) $\dot{\phi}$ about a vertical axis. Thus resolving along the axes of the frame
$$\boldsymbol{\Omega} = (-\dot{\phi}\sin\theta, \dot{\theta}, \dot{\phi}\cos\theta). \tag{3.35}$$
The acceleration of G in an inertial frame can now be calculated, thus:
$$\frac{d\bar{\mathbf{v}}}{dt} = \frac{\partial\bar{\mathbf{v}}}{\partial t} + \boldsymbol{\Omega} \times \bar{\mathbf{v}}$$
$$= (a+b)(\ddot{\theta}, \ddot{\phi}\sin\theta + \dot{\theta}\dot{\phi}\cos\theta, 0)$$
$$+ (a+b)(-\dot{\phi}^2\sin\theta\cos\theta, \dot{\theta}\dot{\phi}\cos\theta, -\dot{\theta}^2 - \dot{\phi}^2\sin^2\theta).$$
Resolving the external forces upon the sphere in the directions of the axes, the equations of motion of G are now found to be
$$m(a+b)(\ddot{\theta} - \dot{\phi}^2\sin\theta\cos\theta) = F_1 + mg\sin\theta, \tag{3.36}$$
$$m(a+b)(\ddot{\phi}\sin\theta + 2\dot{\theta}\dot{\phi}\cos\theta) = F_2, \tag{3.37}$$
$$-m(a+b)(\dot{\theta}^2 + \dot{\phi}^2\sin^2\theta) = R - mg\cos\theta. \tag{3.38}$$
Since the axes of the frame $G123$ are principal axes for the sphere, its angular momentum about G is given by
$$\mathbf{h}_G = mk^2(\omega_1, \omega_2, \omega_3). \tag{3.39}$$
Taking moments of the external forces about G, the equation of angular momentum is
$$(\partial\mathbf{h}_G/\partial t) + \boldsymbol{\Omega} \times \mathbf{h}_G = (F_2 b, -F_1 b, 0).$$
This leads to the equations
$$mk^2(\dot{\omega}_1 + \omega_3\dot{\theta} - \omega_2\dot{\phi}\cos\theta) = F_2 b, \tag{3.40}$$
$$mk^2(\dot{\omega}_2 + \omega_1\dot{\phi}\cos\theta + \omega_3\dot{\phi}\sin\theta) = -F_1 b, \tag{3.41}$$
$$mk^2(\dot{\omega}_3 - \omega_2\dot{\phi}\sin\theta - \omega_1\dot{\theta}) = 0. \tag{3.42}$$
Equations (3.34), (3.36) to (3.38), (3.40) to (3.42), determine the motion. Equations (3.34), (3.42), show that $\omega_3 = n$ (a constant).

Eliminating F_1, F_2 between equations (3.36), (3.37), (3.40), (3.41), and employing the rolling conditions (3.34), we get
$$(k^2 + b^2)(a+b)(\ddot{\phi}\sin\theta + 2\dot{\theta}\dot{\phi}\cos\theta) = k^2 bn\dot{\theta}, \tag{3.43}$$
$$(k^2 + b^2)(a+b)(\ddot{\theta} - \dot{\phi}^2\sin\theta\cos\theta) + k^2 bn\dot{\phi}\sin\theta = gb^2\sin\theta. \tag{3.44}$$

Multiplying equation (3.43) by $\sin \theta$, it integrates immediately to yield

$$(k^2 + b^2)(a + b)\dot{\phi}\sin^2\theta + k^2bn\cos\theta = H, \qquad (3.45)$$

where H is constant. Multiplying equation (3.43) by $\dot{\phi}\sin\theta$ and equation (3.44) by $\dot{\theta}$ and adding, we obtain an equation which also integrates immediately to give

$$\tfrac{1}{2}(k^2 + b^2)(a + b)(\dot{\theta}^2 + \dot{\phi}^2\sin^2\theta) + gb^2\cos\theta = E, \qquad (3.46)$$

where E is the constant of integration. The reader may verify that this is the energy equation.

If equations (3.45), (3.46), are compared with equations (3.17), (3.16), respectively, for the motion of a top, it will be observed that, with $A = m(k^2 + b^2)$, $C = mk^2b/(a+b)$, $l = b^2/(a+b)$, they become identical. The common radius OG therefore always moves like the axis of a top, i.e. it moves between two cones with common vertex O and common axis the vertical through O. This means that the point of contact C moves between a pair of horizontal small circles on the fixed sphere.

In the particular case of steady motion, θ and $\dot{\phi}$ are constant and equations (3.43), (3.44) are satisfied provided

$$(k^2 + b^2)(a + b)\dot{\phi}^2\cos\theta - k^2bn\dot{\phi} + b^2g = 0.$$

This equation possesses real roots in $\dot{\phi}$ if

$$n^2 \geqslant 4g(k^2 + b^2)(a + b)\cos\theta/k^4. \qquad \square$$

EXERCISES

1.   A rigid body whose principal moments of inertia are A, A, C ($C > A$) rotates about its centre of mass under the action of a retarding couple $-K\omega$, where K is constant and ω is its angular velocity. The initial values of the angular velocity components ω_1, ω_2, ω_3 (relative to the principal axes) are $(\Omega_1, 0, \Omega_3)$ respectively. Show that, at a time t later,

$$\omega_1 = \Omega_1 e^{-Kt/A}\cos\lambda, \quad \omega_2 = \Omega_1 e^{-Kt/A}\sin\lambda, \quad \omega_3 = \Omega_3 e^{-Kt/C},$$

where
$$\lambda = \frac{C(C-A)}{KA}\Omega_3(1 - e^{-Kt/C}).$$

2.   A point O is a perpendicular distance c from a smooth plane vertical wall. A uniform rod of length a has one end smoothly pivoted at O and the other end moves on the plane wall. If θ is the inclination of the rod to the horizontal and ϕ is the angle between the vertical plane containing the rod and the vertical plane perpendicular to the wall, show that

$a \cos \theta \cos \phi = c$ and that the energy equation can be written

$$\dot{\theta}^2 \left[1 + \frac{c \sin^2 \theta}{\sqrt{(a^2 \cos^2 \theta - c^2)}} \right] = \frac{3g}{a}(\sin \alpha - \sin \theta),$$

where initially $\theta = \alpha,\ \dot{\theta} = 0$.

3. By differentiating equation (3.18), obtain the following equation for the motion of a top:

$$A(\ddot{\theta} - \dot{\phi}^2 \sin \theta \cos \theta) + Cn\dot{\phi} \sin \theta = Mgl \sin \theta.$$

Putting $\theta = \alpha + \varepsilon,\ \dot{\phi} = \Omega + \eta$, in this equation and equation (3.17) and neglecting second order terms in $\varepsilon,\ \eta$, obtain equations governing small oscillations about the steady state determined by equation (3.15) (with $\theta = \alpha$). Hence show that the frequency of these oscillations is $p/2\pi$, where

$$A^2 p^2 = A^2 \Omega^2 - 2AMgl \cos \alpha + (M^2 g^2 l^2 / \Omega^2).$$

4. If in Problem 3.6, the horizontal plane is constrained to rotate about Oz with angular velocity Ω, but the frame $Oxyz$ remains fixed, show that the equations of linear and angular momentum for the sphere remain unchanged, but (assuming no slipping) the rolling conditions become

$$u - a\omega_2 = -\Omega y, \quad v + a\omega_1 = \Omega x,$$

where (x, y, z) are the coordinates of the centre G of the sphere. Deduce that G describes an ellipse, making one circuit in a time $7\pi/\Omega$.

5. A uniform solid sphere of radius a rolls without slipping inside a fixed circular cylinder of radius b, whose axis is horizontal. If θ is the inclination to the vertical of the plane containing the axis of the cylinder and the centre G of the sphere, u is the component of G's velocity along the cylinder's axis and ω_3 is the component of the sphere's angular velocity along CG, where C is the point of contact of the sphere and cylinder, obtained the equations

$$7(b-a)\dot{\theta}^2 = 10g \cos \theta + \text{const.}, \quad 7u^2 + 2a^2 \omega_3^2 = \text{const.}, \quad \frac{d^2 \omega_3}{d\theta^2} + \frac{2}{7}\omega_3 = 0$$

(Hint: Choose axes $G123$, where $G1$ is parallel to the cylinder's axis and $G3$ is along CG.)

Chapter 4

Equations of Lagrange and Hamilton

4.1 Lagrange's Equations If the configuration of a dynamical system can be specified at any given time t by stating the values of n independent quantities $q_1, q_2, \ldots, q_n$, these are said to be *generalized coordinates* for the system. Thus, for the system described in Problem 2.7 (Fig. 2.6), the distance x moved by the ring at time t from its initial position and the angle θ made by the rod with the vertical, are suitable generalized coordinates.

The kinetic energy of the system T can always be expressed as a function of the q_i and their time derivatives $\dot{q}_i$ (the *generalized velocities*). The dependence of T upon the q_i can take any form, but the kinetic energy is always a quadratic function of the $\dot{q}_i$. T may also depend explicitly upon t if the system's constraints (see below) are variable. Thus, in general,

$$T = T(q_1, \ldots, q_n, \dot{q}_1, \ldots, \dot{q}_n, t) \tag{4.1}$$

The particles of a system are not usually free to execute purely arbitrary motions. These motions are required to satisfy certain geometrical conditions called *constraints*. For example, the two ends of the rod in Fig. 2.3 are constrained to move, one along a vertical line, and the other along a horizontal line; further, since the rod is rigid, any two of its particles must always remain at the same distance from one another. In cases where these geometrical conditions do not change with t, the constraints are said to be *fixed*; in cases where they do depend upon t, the constraints are said to be *variable*, e.g. if in Fig. 2.3 the vertical wall were to be moved in a predetermined manner, this constraint would be variable. If all the constraints are fixed, T will be a *homogeneous* quadratic function of the $\dot{q}_i$.

Lagrange's equations of motion for a system whose kinetic energy has been found in the form (4.1) are

$$\frac{d}{dt}\left(\frac{\partial T}{\partial \dot{q}_i}\right) - \frac{\partial T}{\partial q_i} = Q_i, \tag{4.2}$$

$i = 1, 2, \ldots, n$, where Q_i are the *generalized components of force*. The Q_i are calculated as follows. Suppose the system to be in an arbitrary configuration $(q_1, q_2, \ldots, q_n)$ at time t and let it undergo a small displacement to a neighbouring configuration $(q_1 + \delta q_1, q_2 + \delta q_2, \ldots, q_n + \delta q_n)$. This displacement must be consistent with the constraints, *assumed frozen in in the form they take at the instant t*. Thus, if the constraints are variable, this displacement cannot coincide with the actual motion during the interval

45

$(t, t + \delta t)$; for this reason, the displacement is referred to as a *virtual displacement*. During the virtual displacement, all forces of the system (internal and external) are supposed to remain constant in magnitude and direction. The work done by these forces is always expressible, to the first order, in the form

$$\delta W = \sum_{i=1}^{n} Q_i \, \delta q_i. \tag{4.3}$$

The coefficient of δq_i in this expression accordingly defines the component of force Q_i.

It should be noted that the work done by the internal forces of a rigid body during any virtual displacement which does not disrupt the body, is zero. Similarly, provided the length of a string is not altered in a virtual displacement, its internal tension does no work.

Problem 4.1 Let (r, θ, ϕ) be the usual spherical polar coordinates of a particle m with respect to a rectangular inertial frame. Let (F_r, F_θ, F_ϕ) be the components of the force acting upon the particle in the directions r increasing, θ increasing and ϕ increasing, respectively. Find Lagrange's equations for the motion.

Solution. In a virtual displacement, the particle's coordinates change to $(r + \delta r, \theta + \delta \theta, \phi + \delta \phi)$ and the work done by the force is given by

$$\delta W = F_r \, \delta r + F_\theta r \, \delta \theta + F_\phi r \sin \theta \, \delta \phi.$$

It follows that

$$Q_r = F_r, \quad Q_\theta = r F_\theta, \quad Q_\phi = r \sin \theta \, F_\phi.$$

The spherical polar components of the velocity of the particle are $(\dot{r}, r\dot{\theta}, r \sin \theta \, \dot{\phi})$ and the particle's kinetic energy is accordingly given by

$$T = T(r, \theta, \dot{r}) = \tfrac{1}{2} m (\dot{r}^2 + r^2 \dot{\theta}^2 + r^2 \sin^2 \theta \, \dot{\phi}^2).$$

We can now write down Lagrange's equations associated with the generalized coordinates (r, θ, ϕ),

$$m(\ddot{r} - r\dot{\theta}^2 - r \sin^2 \theta \, \dot{\phi}^2) = F_r,$$

$$m \left[\frac{1}{r} \frac{d}{dt} (r^2 \dot{\theta}) - r \sin \theta \cos \theta \, \dot{\phi}^2 \right] = F_\theta,$$

$$\frac{m}{r \sin \theta} \frac{d}{dt} (r^2 \sin^2 \theta \, \dot{\phi}) = F_\phi.$$

The expressions multiplying m in the left-hand members of these equations are, of course, the spherical polar components of the acceleration vector. $\square$

Problem 4.2 A light string passes over a pulley of mass $4m$ which can rotate freely about a fixed horizontal axis. A particle of mass $2m$ is attached to one end and a pulley of mass m to the other. The second pulley is also free to rotate about a horizontal axis and carries a light string with masses $2m$, $11m$ at its two ends. The system moves in a vertical plane with all strings vertical. Calculate the accelerations of the masses and pulleys.

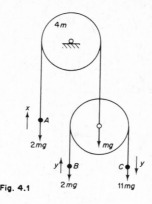

Fig. 4.1

Solution. At time t, suppose the particle A (Fig. 4.1) has moved upwards a distance x and the particles B, C have each moved a distance y relative to the lower pulley. Then x, y are generalized coordinates for the system.

Consider a virtual displacement in which x increases by δx and y by δy. The work done by the weight of A is then $-2mg\,\delta x$ and by the weight of the lower pulley is $mg\,\delta x$. B moves down a distance $\delta x - \delta y$ and its weight does work $2mg(\delta x - \delta y)$. C moves down $\delta x + \delta y$ and its weight does work $11mg(\delta x + \delta y)$. None of the remaining forces of the system do work. Thus

$$\delta W = 12mg\,\delta x + 9mg\,\delta y,$$

whence $\qquad Q_x = 12mg, \; Q_y = 9mg.$

The angular velocity of the lower pulley is $\dot{y}/b$, where b is its radius; the velocity of its centre of mass is $\dot{x}$. Assuming it to be a uniform disc, its kinetic energy is accordingly $\frac{1}{2}m\dot{x}^2 + \frac{1}{4}m\dot{y}^2$. The kinetic energy of the other pulley is $m\dot{x}^2$. The particles B, C have velocities $(\dot{x} - \dot{y})$, $(\dot{x} + \dot{y})$, respectively in the downwards direction. Thus the kinetic energy of the system is

$$
\begin{aligned}
T &= \tfrac{1}{2}m\dot{x}^2 + \tfrac{1}{4}m\dot{y}^2 + m\dot{x}^2 + m(\dot{x} - \dot{y})^2 + \tfrac{11}{2}m(\dot{x} + \dot{y})^2 + m\dot{x}^2 \\
&= \tfrac{9}{4}m(4\dot{x}^2 + 4\dot{x}\dot{y} + 3\dot{y}^2) \\
&= T(\dot{x}, \dot{y}).
\end{aligned}
$$

47

Lagrange's equations can now be written down in the form,

$$\frac{d}{dt}(18\dot{x}+9\dot{y}) = 12g, \qquad \frac{d}{dt}(9\dot{x}+\frac{27}{2}\dot{y}) = 9g.$$

Whence, $\ddot{x} = \frac{1}{2}g$, $\ddot{y} = \frac{1}{3}g$, from which the accelerations of all parts of the system can be calculated. ☐

Problem 4.3 For the system described in Problem 2.7, take x, the distance moved by the ring and θ, the rod's inclination to the vertical, as generalized coordinates.

Solution.

$$T = T(\theta, \dot{x}, \dot{\theta}) = \tfrac{1}{2}m\dot{x}^2 + m(\dot{x}^2 + a^2\dot{\theta}^2 - 2a\dot{\theta}\dot{x}\cos\theta) + \tfrac{1}{3}ma^2\dot{\theta}^2.$$

In the virtual displacement, x increases by δx and θ by $\delta\theta$. Only the weight of the rod does work and, since its centre of mass moves a distance $a\,\delta\theta\sin\theta$ vertically, hence

$$\delta W = -2mga\,\delta\theta\sin\theta.$$

Thus $$Q_x = 0, \qquad Q_\theta = -2mga\sin\theta.$$

Lagrange's equations can now be constructed:

$$\frac{d}{dt}(3\dot{x}-2a\dot{\theta}\cos\theta) = 0, \quad \frac{d}{dt}\left(\frac{8}{3}a^2\dot{\theta}-2a\dot{x}\cos\theta\right)-2a\dot{\theta}\dot{x}\sin\theta = -2ga\sin\theta.$$

Since $\dot{x} = v$, the first of these equations is equivalent to the equation of conservation of horizontal momentum (2.26). The second equation can be written

$$\tfrac{4}{3}a\ddot{\theta} - \ddot{x}\cos\theta + g\sin\theta = 0.$$

After putting $\dot{x} = \frac{2}{3}a\dot{\theta}\cos\theta$, this is found to be identical with the result of differentiating equation (2.28). ☐

4.2 Systems with Variable Constraints

Problem 4.4 A rigid straight rail is fixed in the plane of a lamina and a small engine, of mass m, moves itself by friction drive along the rail. The

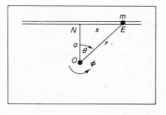

Fig. 4.2

lamina is horizontal and is free to rotate about a vertical axis which does not intersect the rail. Discuss the motion of the system.

Solution. In Fig. 4.2, E is the engine and EN is the rail; O is the point in which the vertical axis of rotation intersects the lamina. $ON = a$ is perpendicular to the rail and $EN = x$. We shall assume that the motion of the engine relative to the rail is known, i.e. x is a known function of t; this constitutes an internal constraint of the system.

Let ϕ be the angle turned through by the lamina; this is the generalized coordinate for the system. If mk^2 is the moment of inertia of the lamina and rail about the axis through O, its kinetic energy is $\frac{1}{2}mk^2\dot{\phi}^2$. The engine's velocity is compounded of its velocity $\dot{x}$ along NE relative to the lamina and the velocity of the lamina at E, namely $r\dot{\phi}$ perpendicular to OE ($r = OE$). Thus, the net kinetic energy of the system is given by

$$T = \tfrac{1}{2}mk^2\dot{\phi}^2 + \tfrac{1}{2}m(\dot{x}^2 + r^2\dot{\phi}^2 - 2r\dot{\phi}\dot{x}\cos\theta),$$

where $\theta = $ angle NOE. Since $r\cos\theta = a$ and $r^2 = x^2 + a^2$, this reduces to

$$T = \tfrac{1}{2}m(x^2 + a^2 + k^2)\dot{\phi}^2 - ma\dot{x}\dot{\phi} + \tfrac{1}{2}m\dot{x}^2.$$

In a virtual displacement, ϕ increases by $\delta\phi$ and the engine remains stationary on its rail. Thus the engine's motor does no work and neither do any of the other forces belonging to the system. Hence $Q_\phi = 0$. Lagrange's equation for the system is now found to be

$$\frac{d}{dt}\left[(x^2 + a^2 + k^2)\dot{\phi} - a\dot{x}\right] = 0.$$

Hence
$$(x^2 + a^2 + k^2)\dot{\phi} = a\dot{x}.$$

This equation determines $\phi(t)$ when $x(t)$ is given. It may, however, be integrated, without any knowledge of $x(t)$, thus:

$$\phi = \int \frac{a\,dx}{x^2 + a^2 + k^2} = \frac{a}{\sqrt{(a^2 + k^2)}}\tan^{-1}\frac{x}{\sqrt{(a^2 + k^2)}},$$

where we have assumed that $\phi = 0$ when $x = 0$. This equation relates the motions of the engine and lamina. In particular, as x varies from $-\infty$ to $+\infty$, it is clear that ϕ varies from $-\pi a/2\sqrt{(a^2 + k^2)}$ to $+\pi a/2\sqrt{(a^2 + k^2)}$ and, hence, that the lamina can never rotate through an angle exceeding $\pi a/\sqrt{(a^2 + k^2)}$. $\qquad\square$

4.3 Conservative Systems If the system is conservative, a potential energy function V exists. V will be expressible in terms of the coordinates q_i and the time t may occur explicitly if the constraints are variable. The work done by the forces of the system in a virtual displacement in which q_i increases by δq_i is given by

$$\delta W = -\delta V = -\sum_{i=1}^{n} \frac{\partial V}{\partial q_i} \delta q_i.$$

Hence
$$Q_i = -\frac{\partial V}{\partial q_i}, \tag{4.4}$$

and Lagrange's equations are

$$\frac{d}{dt}\left(\frac{\partial T}{\partial \dot{q}_i}\right) - \frac{\partial T}{\partial q_i} = -\frac{\partial V}{\partial q_i}. \tag{4.5}$$

Defining a Lagrangian function L by the equation
$$L = L(q_1,\ldots,q_n,\dot{q}_1,\ldots,\dot{q}_n,t) = T - V,$$

Lagrange's equations (4.2) can be expressed in the form

$$\frac{d}{dt}\left(\frac{\partial L}{\partial \dot{q}_i}\right) - \frac{\partial L}{\partial q_i} = 0. \tag{4.6}$$

Problem 4.5 Derive the energy equation from Lagrange's equations in the case of a conservative system with fixed constraints.

Solution. For a conservative system with fixed constraints, T is a homogeneous quadratic function of the generalized velocities. Thus

$$2T = \sum_{i,j} a_{ij}\dot{q}_i\dot{q}_j, \tag{4.7}$$

where the a_{ij} are functions of the coordinates. It follows that

$$\sum_{i} \dot{q}_i \frac{\partial T}{\partial \dot{q}_i} = 2T. \tag{4.8}$$

Multiplying equation (4.5) by $\dot{q}_i$ and summing with respect to i,

$$\sum_{i}\left[\dot{q}_i\frac{d}{dt}\left(\frac{\partial T}{\partial \dot{q}_i}\right) - \dot{q}_i\frac{\partial T}{\partial q_i}\right] = -\sum_{i}\dot{q}_i\frac{\partial V}{\partial q_i}$$

or
$$\sum_{i}\left[\frac{d}{dt}\left(\dot{q}_i\frac{\partial T}{\partial \dot{q}_i}\right) - \ddot{q}_i\frac{\partial T}{\partial \dot{q}_i} - \dot{q}_i\frac{\partial T}{\partial q_i}\right] = -\frac{dV}{dt}, \tag{4.9}$$

since $V = V(q_1,q_2,\ldots,q_n)$. Now
$$T = T(q_1,\ldots,q_n,\dot{q}_1,\ldots,\dot{q}_n),$$

if the constraints are not variable (i.e. T is not explicitly dependent on t) and thus

$$\frac{dT}{dt} = \sum_{i}\left[\dot{q}_i\frac{\partial T}{\partial q_i} + \ddot{q}_i\frac{\partial T}{\partial \dot{q}_i}\right].$$

Using this equation and equation (4.8), equation (4.9) can accordingly be written

$$2\frac{dT}{dt} - \frac{dT}{dt} = -\frac{dV}{dt}$$

and this is equivalent to the energy equation $T + V = $ constant. ◻

Problem 4.6 A uniform rod is smoothly pivoted at one end to a fixed point and moves three dimensionally under gravity. Obtain its equations of motion.

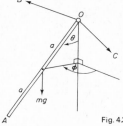

Fig. 4.3

Solution. Let θ be the inclination of the rod to the vertical and ϕ the angle made by the vertical plane containing the rod with a fixed vertical plane. Take axes OA along the rod, OB horizontally and OC to complete a rectangular frame. These are principal axes for the rod at the pivot O and its principal moments of inertia are 0, $\frac{4}{3}ma^2$, $\frac{4}{3}ma^2$, $2a$ being the length of the rod. Its angular velocity about O has components ($\dot{\phi}\cos\theta$, $\dot{\theta}, \dot{\phi}\sin\theta$) and its kinetic energy is accordingly given by

$$2T = \tfrac{4}{3}ma^2(\dot{\theta}^2 + \dot{\phi}^2\sin^2\theta).$$

$V = -mga\cos\theta$ and the Lagrangian is thus

$$L = \tfrac{2}{3}ma^2(\dot{\theta}^2 + \dot{\phi}^2\sin^2\theta) + mga\cos\theta.$$

Lagrange's equations are

$$\ddot{\theta} - \dot{\phi}^2\sin\theta\cos\theta + \frac{3g}{4a}\sin\theta = 0, \quad \frac{d}{dt}(\dot{\phi}\sin^2\theta) = 0.$$

Integrating the second equation, we get

$$\dot{\phi}\sin^2\theta = h,$$

which also follows from the fact that angular momentum about the vertical through O is conserved. The first Lagrange equation can be replaced by the simpler energy equation; however, for the steady motion in which $\theta = \alpha$ (a constant), this equation yields the important result

$$\dot{\phi}^2 = \frac{3g}{4a}\sec\alpha,$$

determining the angular rate $\dot{\phi}$ in terms of the inclination α. ◻

51

Problem 4.7 Find the equations of motion for the two particle system studied in Problem 2.1.

Solution. r and θ are generalized coordinates. The potential energy of the hanging particle can be taken to be $-kmg(l-r)$, where l is the length of the string. Thus, neglecting a constant term,

$$V = kmgr.$$

Also

$$T = \tfrac{1}{2}m(\dot{r}^2 + r^2\dot{\theta}^2) + \tfrac{1}{2}km\dot{r}^2.$$

Thus

$$L = \tfrac{1}{2}m(k+1)\dot{r}^2 + \tfrac{1}{2}mr^2\dot{\theta}^2 - kmgr.$$

Lagrange's equations are

$$\frac{d}{dt}[m(k+1)\dot{r}] - mr\dot{\theta}^2 + kmg = 0, \qquad \frac{d}{dt}(mr^2\dot{\theta}) = 0.$$

The first of these equations follows from equations (2.1), (2.3) by addition and the second is equation (2.2). $\qquad\square$

Problem 4.8 Find the equations of motion for the top studied in Problems 3.4 and 3.5.

Solution. Generalized coordinates for the top are θ, ϕ, ψ, where θ and ϕ are shown in Fig. 3.3 and ψ is the angle between a plane fixed in the top containing the top's axis OA and the vertical plane OAN (also indicated in the figure). The rectangular frame $OABC$ has angular velocity components $(\dot{\phi}\cos\theta, \dot{\theta}, \dot{\phi}\sin\theta)$ in the directions of its axes and the top has angular velocity $(\dot{\psi}, 0, 0)$ relative to this frame. The angular velocity of the top relative to a fixed frame instantaneously coincident with $OABC$ is accordingly $(\dot{\psi} + \dot{\phi}\cos\theta, \dot{\theta}, \dot{\phi}\sin\theta)$. Since the axes of this frame are principal axes of inertia for the top, its kinetic energy is now calculated to be

$$\tfrac{1}{2}A(\dot{\theta}^2 + \dot{\phi}^2\sin^2\theta) + \tfrac{1}{2}C(\dot{\phi}\cos\theta + \dot{\psi})^2.$$

The potential energy of the top is $Mgl\cos\theta$. Hence

$$L = \tfrac{1}{2}A(\dot{\theta}^2 + \dot{\phi}^2\sin^2\theta) + \tfrac{1}{2}C(\dot{\phi}\cos\theta + \dot{\psi})^2 - Mgl\cos\theta$$

and Lagrange's equations take the form

$$\frac{d}{dt}(A\dot{\theta}) - A\dot{\phi}^2\sin\theta\cos\theta + C(\dot{\phi}\cos\theta + \dot{\psi})\dot{\phi}\sin\theta - Mgl\sin\theta = 0.$$

$$\frac{d}{dt}[A\dot{\phi}\sin^2\theta + C(\dot{\phi}\cos\theta + \dot{\psi})\cos\theta] = 0, \qquad \frac{d}{dt}[C(\dot{\phi}\cos\theta + \dot{\psi})] = 0.$$

The last of these equations integrates to give

$$\dot{\phi}\cos\theta + \dot{\psi} = n, \tag{4.10}$$

where, as in Problem 3.4, n is the constant component of the top's angular

52

velocity in the direction of its axis. The second Lagrange equation then integrates to yield

$$A\dot\phi \sin^2\theta + Cn \cos\theta = H. \tag{4.11}$$

This is equation (3.17). The first Lagrange equation can be written

$$A\ddot\theta - A\dot\phi^2 \sin\theta \cos\theta + Cn\dot\phi \sin\theta = Mgl \sin\theta. \tag{4.12}$$

Usually, it is convenient to replace this equation by the energy equation, which has the advantage that it is of first order instead of second order. However, for steady precessional motion $\theta = \alpha$ of the top, equation (4.12) reduces to the important equation (3.15) immediately. ◻

Problem 4.9 Find the equations of motion of the falling square plate of Problem 3.3.

Solution. The kinetic energy of the plate has already been calculated. Its potential energy is $\frac{1}{2}mga \cos\theta$. Hence

$$L = \frac{1}{6}ma^2[\dot\theta^2 + \dot\phi^2(1 + \sin^2\theta) - \frac{3}{2}\dot\theta\dot\phi \cos\theta] - \frac{1}{2}mga \cos\theta.$$

θ, ϕ are, of course, generalized coordinates for this system. Lagrange's equations are now calculated in the form

$$\frac{d}{dt}(\dot\theta - \frac{3}{4}\dot\phi \cos\theta) - \dot\phi^2 \sin\theta \cos\theta - \frac{3}{4}\dot\theta\dot\phi \sin\theta = \frac{3g}{a} \sin\theta,$$

$$\frac{d}{dt}[\dot\phi(1 + \sin^2\theta) - \frac{3}{4}\dot\theta \cos\theta] = 0.$$

The second of these equations is equivalent to equation (3.13). The first equation is best replaced by the simpler energy equation, (3.12). ◻

If the Lagrangian does not contain one of the generalized coordinates explicitly, e.g. q_1, then the associated Lagrange equation simplifies to the form

$$\frac{d}{dt}\left(\frac{\partial L}{\partial \dot q_1}\right) = 0$$

and is immediately integrable to yield

$$\partial L/\partial \dot q_1 = \text{constant}. \tag{4.13}$$

$\partial L/\partial \dot q_1$ is called the *component of generalized momentum* of the system associated with the coordinate q_1 and equation (4.13) can therefore be interpreted as a law of conservation of momentum. As an example, the Lagrangian for a top is independent of both ϕ and ψ. This has the consequence that the associated Lagrange equations integrate to yield equations (4.10), (4.11). It is left for the reader to verify that these equations imply that the components of the angular momentum of the top in the directions of the top's axis and of the vertical are conserved.

4.4 Relaxation of Constraints   If a particular force belonging to a system does no work during a virtual displacement, it will not enter into any of the Lagrange equations and hence cannot be calcufated by this method. Such a force is the reaction P of the smooth wall in problem 2.3. If it is desired to calculate such a force employing the Lagrange technique, it is necessary to imagine that the associated constraint is abolished and that the force is supplied by an alternative external agent. As a consequence, the number of coordinates is increased. Equations of motion for the new system are then written down and these will now include the required force. The manner in which this force must then be chosen in order that the system's motion satisfies the original constraint can now be found. This technique is termed the *method of relaxation of constraints*.

Problem 4.10   Calculate the reaction P of the wall on the rod for the system described in Problem 2.3 (Fig. 2.3).

Solution. Remove the wall. Relaxation of this constraint means that the horizontal motion of the rod is not determined by its rotational motion and the point O beneath the end A of the rod can move horizontally. Let x be the distance of O to the left of a fixed point on the horizontal ground at any time t. Then (x, θ) are generalized coordinates for the relaxed system. Let x be increased by δx and θ by $\delta\theta$ in a virtual displacement. The work done is $\delta W = Mga\,\delta\theta\sin\theta + P\,\delta x$. Thus $Q_\theta = Mga\sin\theta, Q_x = P$.

The centre of mass of the rod has velocity $a\dot\theta$ perpendicular to OG relative to O. This must be added vectorially to the horizontal velocity $\dot x$ of O. The kinetic energy of the rod is therefore given by

$$T = \tfrac{1}{6}Ma^2\dot\theta^2 + \tfrac{1}{2}M(\dot x^2 + a^2\dot\theta^2 + 2a\dot\theta\dot x\cos\theta).$$

Lagrange's equations are accordingly

$$\tfrac{4}{3}a\ddot\theta + \ddot x\cos\theta = g\sin\theta, \qquad M(\ddot x + a\ddot\theta\cos\theta - a\dot\theta^2\sin\theta) = P.$$

We now choose P so that x is constant. The above equations then become

$$\tfrac{4}{3}a\ddot\theta = g\sin\theta, \qquad P = M(a\ddot\theta\cos\theta - a\dot\theta^2\sin\theta).$$

Since $\ddot\theta = d(\tfrac{1}{2}\dot\theta^2)/d\theta$, the first equation can be integrated to yield the energy equation

$$\tfrac{2}{3}a\dot\theta^2 = g(\cos\alpha - \cos\theta)$$

and P now follows exactly as in Problem 2.5.   □

Problem 4.11   Calculate the force exerted on the particle C by the string supporting it in the system of Fig. 4.1.

Solution. We imagine the string cut at C and equal, but opposite, forces F applied to C and to the cut end of the string. Let y denote the distance B has moved upwards relative to the lower pulley and let z denote the

54

distance C has moved downwards with respect to this pulley. Having relaxed the constraint, y and z are now independent coordinates. The kinetic energy is found to be given by

$$T = m[9\dot{x}^2 + \tfrac{5}{4}\dot{y}^2 + \tfrac{11}{2}\dot{z}^2 - 2\dot{x}\dot{y} + 11\dot{x}\dot{z}].$$

In a virtual displacement for which x, y, z increase by δx, δy, δz respectively, the work done by the two forces F is $F(\delta x + \delta y) - F(\delta x + \delta z) = F(\delta y - \delta z)$. Thus, the net work for the whole system is

$$\delta W = F(\delta y - \delta z) - 2mg\,\delta x + mg\,\delta x + 2mg(\delta x - \delta y) + 11mg(\delta x + \delta z)$$

and it follows that

$$Q_x = 12mg, \quad Q_y = F - 2mg, \quad Q_z = -F + 11mg.$$

Lagrange's equations are

$$18\ddot{x} - 2\ddot{y} + 11\ddot{z} = 12g, \quad \frac{5}{2}\ddot{y} - 2\ddot{x} = \frac{F}{m} - 2g, \quad 11\ddot{z} + 11\ddot{x} = -\frac{F}{m} + 11g.$$

Putting $z = y$, these equations are easily solved to give

$$\dot{x} = \tfrac{1}{2}g, \quad \ddot{y} = \tfrac{1}{3}g, \quad F = \tfrac{11}{6}mg.$$

The first two of these equations have already been obtained in Problem 4.2. The third equation giving the force applied to C by the string, can be obtained very simply from the fact that C's acceleration is known to be $\ddot{x} + \ddot{y} = \tfrac{5}{6}g$; its equation of motion is accordingly

$$11mg - F = 11m(\tfrac{5}{6}g). \qquad \square$$

4.5 Non-Holonomic Systems The mechanical systems studied in the previous sections of this chapter have all possessed the property that the increments $\delta q_i (i = 1, 2, \ldots, n)$ in the coordinates determining a virtual displacement could be chosen arbitrarily. Such systems are said to be *holonomic*. This property is crucial for the argument by which Lagrange's equations are established and these equations are accordingly valid for holonomic systems only.

Consider a circular disc of radius a constrained to roll without slipping on a horizontal plane with its plane vertical. Taking axes Oxy parallel to the horizontal plane, let (x, y) be the coordinates of the centre of the disc. Let θ be the angle made by the plane of the disc with the x-axis and let ϕ be the angle through which the disc has rotated. Then (x, y, θ, ϕ) constitute a set of generalized coordinates for the disc. If, however, ϕ is increased by $\delta\phi$, the centre of the disc moves a distance $a\,\delta\phi$ in a direction making an angle θ with Ox and the increments in x and y are determined by the equations

$$\delta x = a\,\delta\phi\cos\theta, \qquad \delta y = a\,\delta\phi\sin\theta. \tag{4.14}$$

This means that the increments δx, δy, $\delta \phi$ are not independent of one another and, hence, cannot be chosen to take arbitrary values. Nevertheless, it is not possible to integrate the equations (4.14) to yield relations between the four coordinates (x, y, θ, ϕ), which could be used to reduce the number of coordinates necessary to specify a configuration of the disc on the plane. For this system, therefore, although there are four coordinates, the increments of only two of these ($\delta\theta$ and $\delta\phi$) can be chosen arbitrarily in a virtual displacement.

In the general case, if $q_1, q_2, \ldots, q_n$ are the generalized coordinates and the virtual increments $\delta q_1, \delta q_2, \ldots, \delta q_n$ are related by r ($< n$) independent conditions, then $n - r$ can be chosen arbitrarily and the system is said to have $n - r$ *degrees of freedom*. For a holonomic system, $r = 0$ and the number of degrees of freedom is equal to the number of generalized coordinates.

Let the r conditions satisfied by the δq_i be

$$\sum_{i=1}^{n} A_{ij} \, \delta q_i = 0, \tag{4.15}$$

where $j = 1, 2, \ldots, r$ and the A_{ij} are functions of the q_i. Then, Lagrange's equations for the system are

$$\frac{d}{dt}\left(\frac{\partial T}{\partial \dot{q}_i}\right) - \frac{\partial T}{\partial q_i} = Q_i + \sum_{j=1}^{r} A_{ij} \lambda_j, \tag{4.16}$$

where $i = 1, 2, \ldots, n$. The r quantities λ_j are called *Lagrange's multipliers*; they are functions of t which are eliminated from the calculation by making use of the r relations between the generalized velocities $\dot{q}_i$ which follow from equations (4.15). For example, if the constraints are fixed, the virtual displacement is a possible actual displacement of the system and hence, dividing the equations (4.15) by δt, we obtain

$$\sum_{i=1}^{n} A_{ij} \, \dot{q}_i = 0 \tag{4.17}$$

as a set of conditions on the generalized velocities.

Problem 4.12   A wheel, of mass m and radius a, rolls without slipping on a horizontal plane. Obtain its equations of motion.

Solution. Oxy are fixed axes in the horizontal plane (Fig. 4.4). The plane of the wheel intersects the horizontal plane in a line l making an angle ϕ with Ox. θ is the inclination of the wheel to the vertical. $C123$ are rectangular axes through the centre of the wheel such that $C1$ passes through its point of contact P with the horizontal plane, $C2$ is parallel to l and $C3$ is perpendicular to the plane of the wheel. Relative to these axes, the wheel has rotated through an angle ψ. Then, if (x, y, z) are the coordinates of C

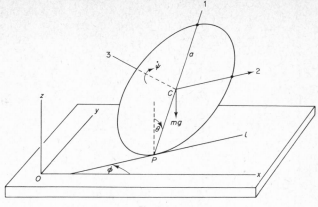

Fig. 4.4

relative to the axes $Oxyz$, the five quantities $(x, y, \theta, \phi, \psi)$ are generalized coordinates for the wheel.

The frame $C123$ has an angular velocity $\dot{\theta}$ about $C2$ and an angular velocity $\dot{\phi}$ about the vertical through C. Thus, the net angular velocity of the frame has components $(\dot{\phi}\cos\theta, \dot{\theta}, \dot{\phi}\sin\theta)$ along its axes. The wheel has angular velocity $\dot{\psi}$ about $C3$ relative to this frame. Hence, the net angular velocity of the wheel has components $(\dot{\phi}\cos\theta, \dot{\theta}, \dot{\psi}+\dot{\phi}\sin\theta)$. The position vector of P relative to C has components $(-a, 0, 0)$. By taking the vector product of these two vectors, we find the velocity of P relative to C, $(0, -a(\dot{\psi}+\dot{\phi}\sin\theta), a\dot{\theta})$. Resolving these components in the directions of the axes $Oxyz$, we get new components

$$-a\dot{\theta}\cos\theta\sin\phi - a(\dot{\psi}+\dot{\phi}\sin\theta)\cos\phi \qquad \text{along } Ox,$$
$$a\dot{\theta}\cos\theta\cos\phi - a(\dot{\psi}+\dot{\phi}\sin\theta)\sin\phi \qquad \text{along } Oy,$$
$$a\dot{\theta}\sin\theta \qquad \text{along } Oz.$$

But C has velocity $(\dot{x}, \dot{y}, \dot{z})$ in the frame $Oxyz$. The components of the net velocity of P in the frame $Oxyz$ can now be found and these must all vanish. Thus, writing $\dot{\psi}+\dot{\phi}\sin\theta = n$,

$$\dot{x} - a\dot{\theta}\cos\theta\sin\phi - an\cos\phi = 0, \qquad (4.18)$$

$$\dot{y} + a\dot{\theta}\cos\theta\cos\phi - an\sin\phi = 0, \qquad (4.19)$$

$$\dot{z} + a\dot{\theta}\sin\theta = 0. \qquad (4.20)$$

These are the rolling conditions at P. Since $z = a\cos\theta$, equation (4.20) is obvious.

57

E

Since the constraints are fixed, a virtual displacement can be an actual displacement and the first two rolling conditions show that the virtual increments δx, etc., are related by the conditions

$$\delta x - a\,\delta\theta\cos\theta\sin\phi - a(\delta\psi + \delta\phi\sin\theta)\cos\phi = 0,$$
$$\delta y + a\,\delta\theta\cos\theta\cos\phi - a(\delta\psi + \delta\phi\sin\theta)\sin\phi = 0.$$

These are the forms taken by equations (4.15) for this problem; the matrix with elements A_{ij} is therefore as set out below:

$$(A_{ij}) = \begin{bmatrix} 1 & 0 \\ 0 & 1 \\ -a\cos\theta\sin\phi & a\cos\theta\cos\phi \\ a\sin\theta\cos\phi & a\sin\theta\sin\phi \\ a\cos\phi & a\sin\phi \end{bmatrix}. \tag{4.21}$$

$C123$ are principal axes for the wheel. Taking the corresponding principal moments of inertia to be $(mk^2, mk^2, 2mk^2)$, the kinetic energy of the rotary motion of the wheel about C can be found using formula (3.11). The Lagrangian then follows in the form

$$L = \tfrac{1}{2}mk^2[\dot\phi^2\cos^2\theta + \dot\theta^2 + 2(\dot\psi + \dot\phi\sin\theta)^2]$$
$$+ \tfrac{1}{2}m(\dot x^2 + \dot y^2 + a^2\dot\theta^2\sin^2\theta) - mga\cos\theta. \tag{4.22}$$

(Note: Use has been made of equation 4.20.)

Lagrange's equations (4.16) associated with the coordinates $(x, y, \theta, \phi, \psi)$ are respectively

$$\frac{d}{dt}(m\dot x) = \lambda_1, \qquad \frac{d}{dt}(m\dot y) = \lambda_2, \tag{4.23}$$

$$\frac{d}{dt}\left[m(k^2 + a^2\sin^2\theta)\dot\theta\right] - mk^2\dot\phi\cos\theta(2n - \dot\phi\sin\theta)$$

$$- ma^2\dot\theta^2\sin\theta\cos\theta - mga\sin\theta = -a\cos\theta(\lambda_1\sin\phi - \lambda_2\cos\phi), \tag{4.24}$$

$$\frac{d}{dt}\left[mk^2(\dot\phi\cos^2\theta + 2n\sin\theta)\right] = -a\sin\theta(\lambda_1\cos\phi + \lambda_2\sin\phi), \tag{4.25}$$

$$2mk^2\dot n = -a(\lambda_1\cos\phi + \lambda_2\sin\phi). \tag{4.26}$$

The first two of these equations show that the multipliers λ_1, λ_2 can be interpreted, physically, as the x- and y-components of the plane's reaction upon the wheel at P. The rather complex third equation may, conveniently, be replaced by the energy equation. These equations, together with the rolling conditions (4.18), (4.19), determine the motion of the wheel.

The general case of the motion of the wheel will not be studied further here (see, however, Exercise 5 at the end of this chapter). Instead, we shall consider two special cases, (i) small oscillations about steady rolling along a straight line and (ii) steady rolling around a circle.

For steady rolling along a straight line, the quantities $\ddot{x}, \ddot{y}, \dot{\phi}, \dot{n}, \theta, \lambda_1, \lambda_2$, all vanish. It is easy to verify that this constitutes a special solution of the equations of motion. If small oscillations are taking place about this motion, all these quantities can be regarded as small. To the first order of small quantities, Lagrange's equations approximate to the forms

$$m\ddot{x} = \lambda_1, \qquad m\ddot{y} = \lambda_2,$$
$$mk^2\ddot{\theta} - 2mk^2 n\dot{\phi} - mga\theta = -a(\lambda_1 \sin \phi - \lambda_2 \cos \phi),$$
$$\ddot{\phi} + 2n\dot{\theta} = 0, \qquad 2mk^2\dot{n} = -a(\lambda_1 \cos \phi + \lambda_2 \sin \phi).$$

Differentiating the rolling conditions (4.18), (4.19), and retaining first order terms only, we find that

$$\ddot{x} = a\ddot{\theta} \sin \phi + a\dot{n} \cos \phi - an\dot{\phi} \sin \phi,$$
$$\ddot{y} = -a\ddot{\theta} \cos \phi + a\dot{n} \sin \phi + an\dot{\phi} \cos \phi.$$

Eliminating $\lambda_1, \lambda_2, \ddot{x}, \ddot{y}$, the following equations are obtained:

$$(k^2+a^2)\ddot{\theta} - (2k^2+a^2)n\dot{\phi} = ga\theta, \qquad \ddot{\phi} + 2n\dot{\theta} = 0, \qquad \dot{n} = 0.$$

Thus n, the angular velocity of the wheel about its axis, can be treated as a constant. Integrating the second equation to give $\dot{\phi} + 2n\theta = 0$ (assume $\dot{\phi} = 0$ when $\theta = 0$) and substituting for $\dot{\phi}$ in the first equation, we get

$$\ddot{\theta} + p^2\theta = 0,$$

where
$$p^2 = \frac{2n^2(2k^2+a^2) - ga}{k^2+a^2}.$$

The condition for small oscillations is now seen to be

$$n^2 > ga/2(2k^2+a^2).$$

In the case of a hoop, $k^2 = \frac{1}{2}a^2$ and the condition is $n^2 > g/4a$. For a uniform disc, $k^2 = \frac{1}{4}a^2$ and the condition is $n^2 > g/3a$.

For steady motion in a circle θ, $\dot{\phi}$ and n are all constant. Thus, the rolling conditions reduce to

$$\dot{x} = an \cos \phi, \qquad \dot{y} = an \sin \phi.$$

Differentiating, $\quad \ddot{x} = -an\dot{\phi} \sin \phi, \qquad \ddot{y} = an\dot{\phi} \cos \phi.$

λ_1, λ_2 can now be calculated and, then, the equation (4.24) gives

$$k^2\dot{\phi} \cos \theta(2n - \dot{\phi} \sin \theta) + ga \sin \theta = -a^2 n\dot{\phi} \cos \theta. \qquad (4.27)$$

59

Suppose the centre of the wheel describes a circle of radius c with angular velocity Ω. Then $\dot{\phi} = -\Omega$ (in Fig. 4.5, ϕ will be decreasing) and $an = c\Omega$. Equation (4.27) accordingly reduces to

$$\Omega^2 = \frac{ga^2 \tan\theta}{k^2(2c + a\sin\theta) + a^2 c}$$

which has already been derived as equation (3.31). □

4.6 Hamilton's Equations   For any conservative system the n quantities $\partial L/\partial \dot{q}_i$ are called the *generalized components of momentum* and are denoted by p_i. Thus

$$p_i = \partial L/\partial \dot{q}_i. \tag{4.28}$$

Since all terms in L are of the second degree at most in the $\dot{q}_i$, p_i depends linearly upon these generalized velocities. It is therefore possible to solve for the $\dot{q}_i$ as linear expressions in the p_i, thus:

$$\dot{q}_i = \sum_{j=1}^{n} B_{ij}p_j + C_i, \tag{4.29}$$

B_{ij}, C_i being functions of the coordinates q_i and t.

Evidently the configuration of the system and the velocities of all its parts, i.e. the *dynamical state* of the system, can be specified by stating the values of the q_i and p_i. q_i and p_i are said to be conjugate quantities.

The Hamiltonian of the system is defined by the equation

$$H = \sum_{i=1}^{n} p_i \dot{q}_i - L, \tag{4.30}$$

where it is assumed that all the velocities $\dot{q}_i$ are replaced by linear expressions in the momenta using equation (4.29). Thus

$$H = H(p_i, q_i, t). \tag{4.31}$$

Problem 4.13   Show that the following equations are equivalent to Lagrange's equations

$$\dot{q}_i = \partial H/\partial p_i, \qquad \dot{p}_i = -\partial H/\partial q_i. \tag{4.32}$$

Solution. From equation (4.30), treating the $\dot{q}_i$ as functions of the basic independent variables (p_i, q_i, t) (equation 4.29), it follows that

$$\frac{\partial H}{\partial q_j} = \sum_i p_i \frac{\partial \dot{q}_i}{\partial q_j} - \frac{\partial L}{\partial q_j} - \sum_i \frac{\partial L}{\partial \dot{q}_i} \frac{\partial \dot{q}_i}{\partial q_j}.$$

Using equation (4.28), this reduces to

$$\frac{\partial H}{\partial q_j} = -\frac{\partial L}{\partial q_j} = -\frac{d}{dt}\left(\frac{\partial L}{\partial \dot{q}_j}\right),$$

where the last equality follows from Lagrange's equations. The second of equations (4.32) has now been proved.

Again from equation (4.30), we have

$$\frac{\partial H}{\partial p_j} = \dot{q}_j + \sum_i p_i \frac{\partial \dot{q}_i}{\partial p_j} - \sum_i \frac{\partial L}{\partial \dot{q}_i} \frac{\partial \dot{q}_i}{\partial p_j} = \dot{q}_j.$$

This is the first of equations (4.32). Equations (4.32) are called *Hamilton's canonical equations* of motion. □

Problem 4.14   Show that, if a system has fixed constraints, then H is the total energy.

Solution. In such a case, the kinetic energy is a homogeneous quadratic form in the $\dot{q}_i$ (equation (4.7)). It follows that

$$p_i = \partial L/\partial \dot{q}_i = \partial T/\partial \dot{q}_i = \sum_j a_{ij} \dot{q}_j.$$

Substituting in equation (4.30), this gives

$$H = \sum_{i,j} a_{ij} \dot{q}_i \dot{q}_j - L = 2T - (T - V) = T + V. \qquad □$$

Problem 4.15   Find Hamilton's equations for a top.

Solution. The Lagrangian for a top has been given in Problem 4.8. The generalized components of momentum are derived from this in the forms

$$p_\theta = \partial L/\partial \dot{\theta} = A\dot{\theta},$$
$$p_\phi = \partial L/\partial \dot{\phi} = A\dot{\phi}\sin^2\theta + C(\dot{\phi}\cos\theta + \dot{\psi})\cos\theta,$$
$$p_\psi = \partial L/\partial \dot{\psi} = C(\dot{\phi}\cos\theta + \dot{\psi}).$$

Solving for the generalized velocities, we obtain the equations

$$\dot{\theta} = p_\theta/A \qquad\qquad\qquad\qquad\qquad\qquad (4.33)$$
$$\dot{\phi} = (p_\phi - p_\psi \cos\theta)/(A\sin^2\theta) \qquad\qquad\quad (4.34)$$
$$\dot{\psi} = -\frac{\cos\theta}{A\sin^2\theta}p_\phi + \left[\frac{1}{C} + \frac{\cos^2\theta}{A\sin^2\theta}\right]p_\psi. \qquad (4.35)$$

Since the constraints are fixed, the Hamiltonian can be identified with the total energy. Thus

$$H = \tfrac{1}{2}A(\dot{\theta}^2 + \dot{\phi}^2\sin^2\theta) + \tfrac{1}{2}C(\dot{\phi}\cos\theta + \dot{\psi})^2 + Mgl\cos\theta$$

$$= \frac{1}{2A}\left[p_\theta^2 + (p_\phi - p_\psi\cos\theta)^2\mathrm{cosec}^2\theta\right] + \frac{1}{2C}p_\psi^2 + Mgl\cos\theta.$$

Hamilton's equations can now be constructed. The first three are simply the equations (4.33)–(4.35); the remainder are as follows:

$$\dot{p}_\theta = \frac{1}{A}(p_\phi - p_\psi\cos\theta)(p_\phi\cos\theta - p_\psi)\mathrm{cosec}^3\theta + Mgl\sin\theta,$$

$$\dot{p}_\phi = \dot{p}_\psi = 0.$$

The first of these equations is easily seen to be equivalent to the Lagrange equation (4.12); the remaining equations integrate to yield equations which are equivalent to equations (4.11) and (4.10). □

Problem 4.16 A bead of mass m is threaded on a smooth circular wire of radius a and the wire is constrained to rotate in its own plane, which is horizontal, about a point O of itself with constant angular velocity ω. Discuss the motion of the bead.

Solution. OA is the diameter of the wire through the pivot O and P is the position of the bead at time t. If C is the centre of the wire, angle $PCA = 2 \times$ angle $POC = 2\theta$. The velocity of the bead relative to a frame rotating with the wire is $2a\dot\theta$ perpendicular to CP. The velocity of this frame at P is $\omega . OP = 2a\omega \cos\theta$ perpendicular to OP. The velocity of the bead in a fixed frame is found by adding these velocities vectorially; thus the kinetic energy of the bead is

$$T = 2ma^2(\dot\theta^2 + \omega^2\cos^2\theta + 2\omega\dot\theta \cos\theta).$$

Since the bead's potential energy is constant, T is also the Lagrangian for the bead. As θ is the only coordinate, the conjugate momentum is

$$p = \partial T/\partial\dot\theta = 4ma^2(\dot\theta + \omega \cos\theta).$$

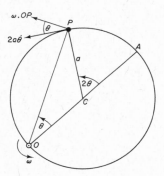

Fig. 4.5

The constraint is variable and H must therefore be calculated from equation (4.30),

$$H = p\dot\theta - 2ma^2(\dot\theta^2 + \omega^2\cos^2\theta + 2\omega\dot\theta \cos\theta) = (p^2/8ma^2) - p\omega \cos\theta.$$

Hamilton's equations are

$$\dot\theta = (p/4ma^2) - \omega \cos\theta, \qquad \dot p = -p\omega \sin\theta.$$

Eliminating p, we calculate that

$$2\ddot\theta + \omega^2\sin 2\theta = 0, \tag{4.36}$$

proving that the angle PCA oscillates about zero in synchrony with a simple pendulum of length g/ω^2. □

Problem 4.17 If $F = F(p_i, q_i, t)$ is a quantity whose value depends on the system's state, prove that

$$\frac{dF}{dt} = [F, H] + \frac{\partial F}{\partial t},$$

where

$$[F, H] = \sum_{i=1}^{n} \left(\frac{\partial F}{\partial q_i} \frac{\partial H}{\partial p_i} - \frac{\partial F}{\partial p_i} \frac{\partial H}{\partial q_i} \right).$$

Solution. We have

$$\frac{dF}{dt} = \frac{\partial F}{\partial q_i} \dot{q}_i + \frac{\partial F}{\partial p_i} \dot{p}_i + \frac{\partial F}{\partial t}$$

and the result now follows immediately by use of Hamilton's equations.

□

$[F, H]$ is called a *Poisson bracket*. In particular, if $F = H$, the Poisson bracket vanishes and we have $dH/dt = \partial H/\partial t$. Thus, if H is not explicitly dependent upon t, it is a constant of the motion. In particular, if the constraints are fixed, this implies that the energy of the system is conserved. However, in the previous example, the energy of the bead is not conserved and yet

$$H = \frac{p^2}{8ma^2} - p\omega \cos\theta = 2ma^2(\dot{\theta}^2 - \omega^2\cos^2\theta) = \text{constant}.$$

The last equation also follows by integration of equation (4.36).

EXERCISES

1. A uniform rod of mass m and length $2a$ has a small light ring attached to one end. The ring is threaded on to a smooth vertical rigid wire and the rod moves in a vertical plane under gravity. If, at time t, the ring is a distance x below a fixed point on the wire and θ is the angle made in the downwards direction by the rod with the horizontal through the ring, obtain Lagrange's equations and deduce that

$$\dot{x} = gt - a\dot{\theta}\cos\theta + \text{constant}, \qquad (4 - 3\cos^2\theta)\dot{\theta}^2 = \text{constant}.$$

2. Four uniform rods, each of mass m and length $2a$, are smoothly joined together at their ends to form a rhombus $ABCD$. AB is smoothly pivoted at its centre and the configuration moves in a vertical plane with

DC below AB. If θ is the angle made by AB and DC with the horizontal and ϕ is the angle made by AD and BC with the vertical, show that the Lagrangian for the system is given by

$$L = \tfrac{2}{3}ma^2(2\dot\theta^2 + 5\dot\phi^2) + 4mga\cos\phi.$$

Deduce that AB and DC rotate with constant angular velocity and that AD and BC oscillate in synchrony with a simple pendulum of length $5a/3$.

3.   The ends of a uniform rod of mass $3m$ and length $2a$ are constrained to move along two smooth wires OA and OB, OA being horizontal and OB vertically downwards. A smooth bead of mass m is free to slide on the rod. The whole system is made to rotate about OB with constant angular velocity $\sqrt{(6g/5a)}$. If θ is the inclination of the rod to OA, obtain Lagrange's equations and show that these possess a solution corresponding to steady motion with the bead at rest at the mid-point of the wire and $\sin\theta = \tfrac{2}{3}$.

4. AC, BC are two equal uniform bars, smoothly hinged at C. A light ring attached to A is threaded on a smooth horizontal wire Ox and a similar ring attached to B moves on a smooth vertical wire Oy, pointing upwards. The bars move in the vertical plane Oxy with C above Ox. If θ, ϕ are the respective inclinations of AC, BC to the horizontal (measured in the same sense), calculate Lagrange's equations satisfied by these coordinates. If, initially, $\theta = \tfrac{1}{3}\pi$, $\phi = \tfrac{1}{6}\pi$, show that, at this instant, the angular accelerations of the bars are $9g/11a$ and $3\sqrt{3}g/22a$.

5.   In the problem of the wheel rolling on a horizontal plane (Problem 4.12), eliminate x, y, λ_1, λ_2 from the Lagrange equations (4.23)–(4.26) and the rolling conditions (4.18), (4.19), and hence obtain the equations

$$(k^2 + a^2)\ddot\theta - (2k^2 + a^2)n\dot\phi\cos\theta + k^2\dot\phi^2\sin\theta\cos\theta = ga\sin\theta,$$
$$\ddot\phi\cos\theta - 2\dot\theta\dot\phi\sin\theta + 2n\dot\theta = 0, \qquad (2k^2 + a^2)\dot n + a^2\dot\theta\dot\phi\cos\theta = 0.$$

Deduce that n satisfies the equation

$$(2k^2 + a^2)\frac{d}{d\mu}\left[(1 - \mu^2)\frac{dn}{d\mu}\right] = 2a^2n,$$

where $\mu = \sin\theta$.

6.   Approximate the equations for a rolling wheel obtained in the previous exercise in the case where θ and n are small and $\dot\phi = \omega + \varepsilon$, where ω is constant and ε is small. Hence show that a wheel spinning with angular velocity ω about a vertical diameter is in stable motion provided $\omega^2 > ga/(k^2 + a^2)$.

7.   A uniform circular wire of radius a and mass $2m$ is free to rotate about

a fixed vertical diameter. A smooth bead of mass m is threaded on the wire. Show that the Hamiltonian for the system is

$$H = \frac{1}{2ma^2}\left\{p_\theta^2 + \frac{p_\phi^2}{1+\sin^2\theta}\right\} - mga\cos\theta,$$

where ϕ is the angle through which the wire has rotated and θ is the angle made by the radius to the bead with the downward vertical. Obtain Hamilton's equations and derive from these the equations

$$\dot\phi(1+\sin^2\theta) = \text{constant}, \qquad \ddot\theta - \dot\phi^2\sin\theta\cos\theta = -g\sin\theta/a.$$

Show that a steady motion is possible with $\dot\phi = \omega, \theta = \alpha$ and $a\omega^2 = g\sec\alpha$.

Chapter 5

Small Oscillations about Equilibrium

5.1 Stability of Equilibrium   Consider a holonomic conservative system whose constraints are fixed. If there exists a configuration in which it can rest in equilibrium, then its equations of motion (4.5) must possess a solution in which the coordinates q_i are all constant and the velocities $\dot{q}_i$ all vanish. Since the constraints are fixed, T will be given by equation (4.7) and it follows that

$$\partial T/\partial \dot{q}_i = \partial T/\partial q_i = 0$$

identically for this particular solution. Substituting in equation (4.5), we deduce that

$$\partial V/\partial q_i = 0, \tag{5.1}$$

$i = 1, 2, \ldots, n$, in the position of equilibrium; i.e. V is stationary in such a position.

If the position of equilibrium is *stable*, then V is a relative minimum in this position; for otherwise it will be possible for the system, after a small disturbance, to move away from the position in such a way as to decrease V and, by the energy equation, this will lead to an increase in T and thus to a more rapid movement of the system away from equilibrium.

Problem 5.1   A uniform rod AB of length $2a$ and mass m has its end A connected to a point C of a smooth vertical wall by an elastic string of natural length $\frac{1}{2}a\sqrt{7}$ and Hooke's constant mg/a. The end B of the rod rests against the wall vertically below C. Show that there is only one equilibrium configuration and that it is unstable.

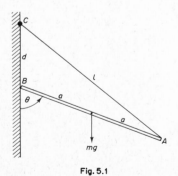

Fig. 5.1

Solution. Consider a configuration in which B is a distance d below C and the length of the string is l. Then, if θ is the angle indicated in Fig. 5.1, by the cosine rule

$$\cos \theta = (l^2 - 4a^2 - d^2)/4ad. \qquad (5.2)$$

The potential energy of a string whose Hooke's constant is k and whose extension is x, is $\frac{1}{2}kx^2$. Thus the potential energy of our configuration is given by

$$V = -mg(d + a\cos\theta) + \frac{mg}{2a}(l - \tfrac{1}{2}a\sqrt{7})^2$$

$$= -\frac{mg}{4d}(3d^2 + l^2 - 4a^2) + \frac{mg}{2a}(l - \tfrac{1}{2}a\sqrt{7})^2,$$

having substituted for $\cos\theta$ from equation (5.2).

We now introduce dimensionless variables x, y by putting $d = ax$ and $l = ay$. Then

$$V = \tfrac{1}{4}mga\left[\frac{4 - y^2}{x} - 3x + 2(y - \tfrac{1}{2}\sqrt{7})^2\right]. \qquad (5.3)$$

For equilibrium, V must be stationary with respect to variations of x and y. The conditions for this are

$$\frac{\partial V}{\partial x} = \tfrac{1}{4}mga\left[\frac{y^2 - 4}{x^2} - 3\right] = 0, \qquad (5.4)$$

$$\frac{\partial V}{\partial y} = \tfrac{1}{4}mga\left[-\frac{2y}{x} + 4(y - \tfrac{1}{2}\sqrt{7})\right] = 0. \qquad (5.5)$$

Equation (5.5) shows that

$$y = x\sqrt{7}/(2x - 1). \qquad (5.6)$$

Substituting from this equation for y into equation (5.4), we obtain

$$3x^4 - 3x^3 + 3x^2 - 4x + 1 = 0.$$

By inspection, $x = 1$ is seen to be a root and taking out the factor $(x - 1)$, we get

$$(x - 1)(3x^3 + 3x - 1) = 0.$$

If $f(x) = 3x^3 + 3x - 1$, then $f'(x) = 9x^2 + 3 > 0$ and hence $f(x)$ increases monotonically. It follows that $f(x)$ can have but one zero and, since $f(0) = -1$, $f(\frac{1}{2}) = \frac{7}{8}$, this lies in the interval $(0, \frac{1}{2})$. But for values of x in this interval equation (5.6) shows that y is negative and this is impossible.

Thus the only configuration for which V is stationary is the one determined by $x = 1$, $y = \sqrt{7}$, i.e. $d = a$, $l = a\sqrt{7}$, $\theta = \tfrac{1}{3}\pi$. In this configuration

67

$$\partial^2 V/\partial x^2 = mga(4-y^2)/2x^3 = -3mga/2,$$
$$\partial^2 V/\partial y^2 = mga(2x-1)/2x = mga/2,$$
$$\partial^2 V/\partial x \partial y = mgay/2x^2 = \sqrt{7}mga/2.$$

Thus
$$\left[\frac{\partial^2 V}{\partial x \partial y}\right]^2 - \frac{\partial^2 V}{\partial x^2}\frac{\partial^2 V}{\partial y^2} = \tfrac{5}{2}mga > 0$$

and the stationary value of V is neither a maximum nor a minimum. It follows that the equilibrium is unstable. □

Problem 5.2   Equal point electric charges are fixed at the vertices of a square. A similar charge is constrained to move in the plane of the square. Show that it can rest in stable equilibrium at the centre of the square.

Solution. Taking axes Oxy through the square's centre parallel to its sides, the coordinates of the vertices can be taken to be $(\pm a, \pm a)$. Let (x, y) be the coordinates of the movable charge when it is displaced slightly from O. The potential energy of a movable charge e when at a distance r from an equal fixed charge is $e^2/r(e^2/4\pi\varepsilon_0$ if S.I. units are used). Thus the potential energy of e at (x, y) in the field of the equal charge at (a, a) is

$$e^2[(x-a)^2+(y-a)^2]^{-\frac{1}{2}} = \frac{e^2}{\sqrt{2a}}\left[1-\frac{1}{a}(x+y)+\frac{1}{2a^2}(x^2+y^2)\right]^{-\frac{1}{2}}$$
$$= \frac{e^2}{\sqrt{2a}}\left[1+\frac{1}{2a}(x+y)+\frac{1}{8a^2}(x^2+y^2+6xy)\right],$$

to the second order in the small quantities x and y. Three similar expressions can be derived for the contributions of the charges at the other three vertices. The net potential energy of the movable charge is calculated to be given by

$$V = \frac{e^2}{2\sqrt{2a^3}}(8a^2+x^2+y^2).$$

It is now evident that V is a minimum at O and hence that the movable charge is in stable equilibrium at this point.

5.2 Small Oscillations about Stable Equilibrium   If a holonomic conservative system with fixed constraints is slightly disturbed from a position of stable equilibrium, it will commence to oscillate about this position. We shall choose the generalized coordinates q_i in such a way that these are all zero in the equilibrium position and shall assume that, together with the velocities $\dot{q}_i$, they remain small during the oscillatory motion. By neglecting terms of order higher than the second in the Lagrangian, an approximate solution for the oscillatory motion can be found.

Problem 5.3 A light string is stretched between two fixed points A, D on a smooth horizontal table with tension P. Two particles B, C of mass m are fixed to the string at the points of trisection. Discuss possible oscillatory motions of the particles in the plane of the table, at right angles to the string.

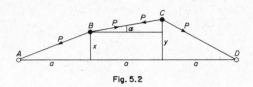

Fig. 5.2

Solution. Let x, y be displacements of the particles from their equilibrium positions (Fig. 5.2). If x, y are small, the tensions in the three segments of the string will remain P to the first order. Consider the component in the direction of the displacement x of the tension in BC acting upon the particle B; this is $P \sin \alpha \doteq P \tan \alpha = P(y-x)/a$, where $AD = 3a$. Similarly, the same component of the tension in BA is $-Px/a$. The net force acting upon B in the direction of the displacement x is accordingly $P(y-2x)/a$. Similarly, the net force acting upon C in the direction of y is $P(x-2y)/a$. Hence, if x increases by δx and y by δy, the work done by the forces of the system is

$$P[(y-2x)\delta x + (x-2y)\delta y]/a$$

and it follows that the generalized forces for the system are given by

$$Q_x = P(y-2x)/a, \qquad Q_y = P(x-2y)/a.$$

The kinetic energy of the system is given by

$$T = \tfrac{1}{2}m(\dot{x}^2 + \dot{y}^2)$$

and Lagrange's equations are therefore

$$\frac{d}{dt}(m\dot{x}) = \frac{P}{a}(y-2x), \qquad \frac{d}{dt}(m\dot{y}) = \frac{P}{a}(x-2y).$$

Putting $\omega^2 = P/ma$, these can be written

$$\ddot{x} + \omega^2(2x-y) = \ddot{y} + \omega^2(2y-x) = 0. \tag{5.7}$$

It is physically evident that these equations possess an oscillatory solution and we accordingly seek a solution in the form

$$x = A\sin(pt+\alpha), \qquad y = B\sin(pt+\alpha).$$

Upon substitution, it is found that the equations are satisfied provided

$$(p^2 - 2\omega^2)A + \omega^2 B = 0, \tag{5.8}$$

$$\omega^2 A + (p^2 - 2\omega^2)B = 0. \tag{5.9}$$

69

For these equations to have solutions other than $A = B = 0$ (the equilibrium solution), it is necessary that

$$\begin{vmatrix} p^2 - 2\omega^2 & \omega^2 \\ \omega^2 & p^2 - 2\omega^2 \end{vmatrix} = 0,$$

i.e. $(p^2 - \omega^2)(p^2 - 3\omega^2) = 0.$ \hfill (5.10)

Thus, $p = \omega$ or $\sqrt{3}\omega$ (negative values of p do not provide distinct solutions).

In the case $p = \omega$, either of the equations (5.8), (5.9), show that $A = B$ and the solution found is

$$x = y = A\sin(\omega t + \alpha),$$ \hfill (5.11)

A and α being arbitrary. The configuration of the particles in this type of motion is shown in Fig. 5.3(a); the particles oscillate in phase with one another with period $2\pi/\omega$.

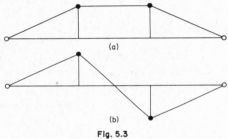

(a)

(b)

Fig. 5.3

In the case $p = \sqrt{3}\omega$, equations (5.8), (5.9), show that $A = -B$ and the motion is determined by

$$x = -y = A\sin(\sqrt{3}\omega t + \alpha).$$ \hfill (5.12)

This type of motion is illustrated in Fig. 5.3(b); the particles oscillate in anti-phase with one another with period $2\pi/\sqrt{3}\omega$.

These two possible types of oscillatory motion are called the *normal modes* of vibration of the system. The characteristic feature of a normal mode is that all particles of the system oscillate in phase or in antiphase with one another.

Since the equations (5.7) governing the system's behaviour are linear, it follows that the solutions (5.11), (5.12) may be added to yield a more general solution,

$$\left.\begin{array}{l} x = A\sin(\omega t + \alpha) + A'\sin(\sqrt{3}\omega t + \alpha'), \\ y = A\sin(\omega t + \alpha) - A'\sin(\sqrt{3}\omega t + \alpha'). \end{array}\right\}$$ \hfill (5.13)

This solution contains four arbitrary constants A, A', α, α' and is the general solution of the equations (5.7). Given the initial positions of the particles

70

and their velocities, these four constants can be found and the subsequent motion determined. For example, if the system is started from rest at $t = 0$ with $x = a$, $y = -3a$, then $A = a$, $A' = 2a$, $\alpha = 3\pi/2$, $\alpha' = \pi/2$, i.e.

$$x = a(-\cos \omega t + 2 \cos \sqrt{3}\omega t), \qquad y = -a(\cos \omega t + 2 \cos \sqrt{3}\omega t). \qquad \square$$

Problem 5.4 Two uniform rods having the same length $2a$ and masses $2m$, $3m$, are smoothly hinged together at one end of each. The free end of the less massive rod is smoothly hinged to a fixed point, so that the rods are free to swing in a vertical plane under gravity. Calculate the periods of the normal modes of small oscillations.

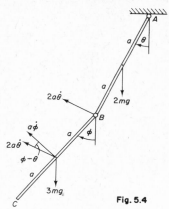

Fig. 5.4

Solution. Let θ, ϕ be the inclinations of the rods AB, BC to the vertical at time t (Fig. 5.4). The velocity of the pivot B is $2a\dot\theta$ and the velocity of the centre of mass of BC relative to B is $a\dot\phi$. The net velocity v of the centre of mass of BC is obtained by summing these velocities vectorially; we find that

$$v^2 = a^2[4\dot\theta^2 + \dot\phi^2 + 4\dot\theta\dot\phi \cos(\phi - \theta)].$$

An expression for the kinetic energy of the system can now be constructed,

$$T = \tfrac{4}{3}ma^2\dot\theta^2 + \tfrac{1}{2}ma^2\dot\phi^2 + \tfrac{3}{2}mv^2.$$

Substituting for v^2 and subtracting the potential energy, we generate the Lagrangian,

$$L = ma^2[22\dot\theta^2/3 + 2\dot\phi^2 + 6\dot\theta\dot\phi \cos(\phi - \theta)]$$
$$+ 2mga \cos \theta + 3mga(2 \cos \theta + \cos \phi).$$

We next approximate this expression to the second order of small quantities θ, ϕ, $\dot\theta$, $\dot\phi$. The result is

$$L = ma^2(\tfrac{22}{3}\dot\theta^2 + 2\dot\phi^2 + 6\dot\theta\dot\phi) - mga(4\theta^2 + \tfrac{3}{2}\phi^2). \qquad (5.14)$$

71

Lagrange's equations for the oscillations are now found to be

$$\frac{44}{3}\ddot{\theta}+6\ddot{\phi}+\frac{8g}{a}\theta = 0, \qquad 6\ddot{\theta}+4\ddot{\phi}+\frac{3g}{a}\phi = 0.$$

Taking a normal mode solution in the form

$$\theta = A\sin(pt+\alpha), \qquad \phi = B\sin(pt+\alpha),$$

we find after substitution that A, B must satisfy the equations

$$\left[\frac{8g}{a}-\frac{44}{3}p^2\right]A-6p^2B = 0, \qquad -6p^2A+\left[\frac{3g}{a}-4p^2\right]B = 0. \qquad (5.15)$$

For a non-null solution, it is necessary that

$$\left[\frac{8g}{a}-\frac{44}{3}p^2\right]\left[\frac{3g}{a}-4p^2\right]-36p^4 = 0.$$

Hence $p^2 = 3g/a$ or $6g/17a$ and the periods of the normal modes are accordingly

$$2\pi\sqrt{(a/3g)}, \qquad 2\pi\sqrt{(17a/6g)}.$$

If $p^2 = 3g/a = \omega_1^2$, equations (5.15) have the solution $A:B = 1:-2$ and the normal mode oscillation is given by

$$\theta = a\sin(\omega_1 t+\alpha), \qquad \phi = -2a\sin(\omega_1 t+\alpha).$$

If $p^2 = 6g/17a = \omega_2^2$, then $A:B = 3:4$ and the normal mode is

$$\theta = 3a\sin(\omega_2 t+\alpha), \qquad \phi = 4a\sin(\omega_2 t+\alpha).$$

The general solution is obtained by superposition of these normal modes,

$$\left.\begin{array}{l} \theta = a\sin(\omega_1 t+\alpha)+3a'\sin(\omega_2 t+\alpha'), \\ \phi = -2a\sin(\omega_1 t+\alpha)+4a'\sin(\omega_2 t+\alpha'). \end{array}\right\} \qquad (5.16)$$

□

5.3 Normal Coordinates
Problem 5.5   Calculate the normal coordinates of the double pendulum studied in the previous problem.

Solution. Referring to equations (5.16), it is evident that if we introduce a new pair of generalized coordinates x, y for the system, related to the old pair θ, ϕ by the linear transformation equations

$$\theta = x+3y, \qquad \phi = -2x+4y, \qquad (5.17)$$

then the general oscillatory motion of the system is expressed by the equations

$$x = a\sin(\omega_1 t+\alpha), \qquad y = a'\sin(\omega_2 t+\alpha'). \qquad (5.18)$$

In particular, in the first normal mode, $a' = 0$ and, hence, y is identically zero; in the second normal mode $a = 0$ and x is identically zero.

Further, if we substitute from equations (5.17) into the Lagrangian (5.14), it reduces to the form

$$L = \tfrac{10}{3}ma^2(\dot{x}^2 + 51\dot{y}^2) - 10mga(x^2 + 6y^2), \qquad (5.19)$$

i.e. both quadratic forms representing the kinetic and potential energies of the system have been reduced to a sum of squares (*diagonal form*).

The simplification can be carried a little further by introducing coordinates u, v, related to x, y by the equations

$$u = \sqrt{(\tfrac{20}{3}ma^2x)}, \qquad v = \sqrt{(340ma^2y)}. \qquad (5.20)$$

In terms of u and v, the Lagrangian (5.19) then assumes the form

$$L = \tfrac{1}{2}(\dot{u}^2 + \dot{v}^2) - g(\tfrac{3}{2}u^2 + \tfrac{3}{17}v^2)/a = \tfrac{1}{2}(\dot{u}^2 + \dot{v}^2 - \omega_1^2 u^2 - \omega_2^2 v^2). \quad (5.21)$$

Thus, if the coordinates u, v are used, the kinetic and potential energies are expressed in the canonical forms

$$T = \tfrac{1}{2}(\dot{u}^2 + \dot{v}^2), \qquad V = \tfrac{1}{2}(\omega_1^2 u^2 + \omega_2^2 v^2). \qquad (5.22)$$

Coordinates which permit the kinetic and potential energies of an oscillatory system to be written in these forms are termed *normal coordinates*. The corresponding Lagrange equations are clearly

$$\ddot{u} + \omega_1^2 u = 0, \qquad \ddot{v} + \omega_2^2 v = 0. \qquad (5.23)$$

One normal mode is now specified by the equations

$$u = P\sin(\omega_1 t + \alpha), \qquad v = 0, \qquad (5.24)$$

and the other normal mode by the equations

$$u = 0, \qquad v = P'\sin(\omega_2 t + \alpha'). \qquad (5.25)$$

This illustrates another characteristic of normal coordinates, that in a normal mode, all but one remain zero.

To summarize, for the double pendulum, the normal coordinates are related to the angles θ, ϕ by the equations

$$\theta\sqrt{(340ma^2)} = u\sqrt{(51)} + 3v, \quad \phi\sqrt{(340ma^2)} = -2u\sqrt{(51)} + 4v,$$
$$\text{i.e. } u = (4\theta - 3\phi)\sqrt{(ma^2/15)}, \quad v = (2\theta + \phi)\sqrt{(17ma^2/5)}. \quad (5.26) \quad \square$$

In general, approximating to the second order of small quantities q_i, $\dot{q}_i$, the kinetic and potential energies of a system can be written in the forms

$$T = \tfrac{1}{2}\sum_{i,j} a_{ij}\dot{q}_i\dot{q}_j, \qquad V = \tfrac{1}{2}\sum_{i,j} b_{ij}q_iq_j, \qquad (5.27)$$

where a_{ij}, b_{ij} are constants. Both these quadratic forms will be *positive definite*, i.e. will be greater than zero for all sets of values of the q_i, $\dot{q}_i$ (except that $T = V = 0$ for the null set). It is proved in textbooks of algebra that

73

two such positive definite quadratic forms can be reduced to the diagonal forms

$$T = \tfrac{1}{2} \sum_i \dot{u}_1^2, \qquad V = \tfrac{1}{2} \sum_i \omega_i^2 u_i^2, \tag{5.28}$$

by a real regular linear transformation (from the coordinates q_i to new coordinates u_i) of the form

$$u_i = \sum_j c_{ij} q_j, \tag{5.29}$$

where the c_{ij} are constants. u_i are the *normal coordinates* for the system.

With T and V in the forms (5.28), Lagrange's equations for the system are

$$\ddot{u}_i + \omega_i^2 u_i = 0, \tag{5.30}$$

$i = 1, 2, \ldots, n$. Substituting the normal mode solution

$$u_i = A_i \sin(pt + \alpha), \tag{5.31}$$

we obtain the equations

$$A_i(\omega_i^2 - p^2) = 0, \tag{5.32}$$

$i = 1, 2, \ldots, n$. Thus, $p = \omega_i$ or $A_i = 0$. If we take $p = \omega_1$, then $A_2, A_3, \ldots, A_n$ must all vanish (we here assume that the ω_i are all different). Thus, the first normal mode is

$$u_1 = A_1 \sin(\omega_1 t + \alpha), \quad u_2 = u_3 = \ldots = u_n = 0. \tag{5.33}$$

Similarly, in the second normal mode only u_2 is non-zero, and so on. We conclude that, in any normal mode, all the normal coordinates vanish except one which oscillates sinusoidally.

Problem 5.6 Two equal uniform straight beams AB, BC, of mass m and length $2a$, are smoothly hinged together at B. They are supported at A, B and C by light vertical springs such that, in equilibrium, A, B, C are at the same level. The stiffness of the spring supporting B is double that of the springs at A and C. Calculate the normal modes of vertical oscillation of the beams and determine the normal coordinates of the system.

Solution. Let k be the Hooke's constant for the springs at A and C, and $2k$ the constant of the spring at B. In equilibrium, the springs at A, C will be subjected to compressive forces $\tfrac{1}{2}mg$ and will each contract a distance $mg/2k$. The central spring will support a weight mg and will also contract a distance $mg/2k$. At some instant t during motion, let A, B, C have moved distances x, y, z vertically upwards respectively from their equilibrium positions. The potential energies of the springs in this configuration will be $\tfrac{1}{2}k(x - mg/2k)^2$, $k(y - mg/2k)^2$, $\tfrac{1}{2}k(z - mg/2k)^2$, respectively. The centres of the beams will have been raised distances $\tfrac{1}{2}(x + y)$, $\tfrac{1}{2}(y + z)$ above their equilibrium positions. It follows that the potential energies of the weights

74

are $\frac{1}{2}mg(x+y)$, $\frac{1}{2}mg(y+z)$. The net potential energy of the system is accordingly.

$$V = \frac{k}{2}\left[x - \frac{mg}{2k}\right]^2 + k\left[y - \frac{mg}{2k}\right]^2 + \frac{k}{2}\left[z - \frac{mg}{2k}\right]^2 + \frac{1}{2}mg\,(x+2y+z),$$

$$= \tfrac{1}{2}k(x^2 + 2y^2 + z^2) + \text{constant.} \tag{5.34}$$

The velocities of the centres of mass of the beams are $\frac{1}{2}(\dot x + \dot y)$, $\frac{1}{2}(\dot y + \dot z)$. The velocity of A relative to B is $\dot x - \dot y$ and it follows that the angular velocity of AB is $(\dot x - \dot y)/2a$. Similarly, the angular velocity of BC is $(\dot y - \dot z)/2a$. We can now write down the kinetic energy of the beams,

$$T = \tfrac{1}{8}m(\dot x + \dot y)^2 + \tfrac{1}{8}m(\dot y + \dot z)^2 + \tfrac{1}{24}m(\dot x - \dot y)^2 + \tfrac{1}{24}m(\dot y - \dot z)^2,$$

$$= \tfrac{1}{6}m(\dot x^2 + 2\dot y^2 + \dot z^2 + \dot x\dot y + \dot y\dot z). \tag{5.35}$$

Forming Lagrange's equations, we get

$$2\ddot x + \ddot y + \omega^2 x = 0, \quad \ddot x + 4\ddot y + \ddot z + 2\omega^2 y = 0, \quad \ddot y + 2\ddot z + \omega^2 z = 0,$$

where $\omega^2 = 6k/m$. Substituting the normal mode solution

$$x = A\sin(pt + \alpha), \quad y = B\sin(pt + \alpha), \quad z = C\sin(pt + \alpha),$$

the equations are satisfied provided

$$\left.\begin{array}{r}(\omega^2 - 2p^2)A - p^2 B = 0, \\ -p^2 A + (2\omega^2 - 4p^2)B - p^2 C = 0, \\ -p^2 B + (\omega^2 - 2p^2)C = 0.\end{array}\right\} \tag{5.36}$$

For a non-null solution, it is necessary that

$$\begin{vmatrix} \omega^2 - 2p^2 & -p^2 & 0 \\ -p^2 & 2\omega^2 - 4p^2 & -p^2 \\ 0 & -p^2 & \omega^2 - 2p^2 \end{vmatrix} = 0.$$

Thus, $p = \omega/\sqrt{3}$, $\omega/\sqrt{2}$ or ω. If $p = \omega/\sqrt{3}$, equations (5.36) show that $A:B:C = 1:1:1$ and the normal mode is given by

$$x = y = z = a\sin[(\omega t/\sqrt{3}) + \alpha].$$

In this motion, the beams continue in a horizontal straight line as they oscillate. If $p = \omega/\sqrt{2}$, then $A:B:C = 1:0:-1$ and the normal mode is

$$x = -z = b\sin[(\omega t/\sqrt{2}) + \beta], \quad y = 0.$$

In this motion, the points A, C oscillate with equal amplitudes in antiphase and B remains stationary. If $p = \omega$, $A:B:C = 1:-1:1$. The normal mode is

$$x = -y = z = c\sin(\omega t + \gamma).$$

A, B and C now all oscillate with equal amplitudes, but B is in anti-phase with A and C.

75

The general solution of the equations of motion is given by

$$x = a\sin\left(\frac{1}{\sqrt{3}}\omega t+\alpha\right)+b\sin\left(\frac{1}{\sqrt{2}}\omega t+\beta\right)+c\sin(\omega t+\gamma),$$

$$y = a\sin\left(\frac{1}{\sqrt{3}}\omega t+\alpha\right)-c\sin(\omega t+\gamma).$$

$$z = a\sin\left(\frac{1}{\sqrt{3}}\omega t+\alpha\right)-b\sin\left(\frac{1}{\sqrt{2}}\omega t+\beta\right)+c\sin(\omega t+\gamma).$$

These equations suggest a transformation to new coordinates ξ, η, ζ, where

$$x = \xi+\eta+\zeta, \quad y = \xi-\zeta, \quad z = \xi-\eta+\zeta.$$

Substituting from these equations into equations (5.34), (5.35), we find that

$$V = \tfrac{1}{2}k(4\xi^2+2\eta^2+4\zeta^2), \quad T = \tfrac{1}{6}m(6\dot{\xi}^2+2\dot{\eta}^2+2\dot{\zeta}^2).$$

Normal coordinates u, v, w can now be constructed by taking

$$u = \sqrt{(2m)}\xi, \quad v = \sqrt{(\tfrac{2}{3}m)}\eta, \quad w = \sqrt{(\tfrac{2}{3}m)}\zeta.$$

Then $\quad V = \tfrac{1}{2}(\tfrac{1}{3}\omega^2u^2+\tfrac{1}{2}\omega^2v^2+\omega^2w^2), \quad T = \tfrac{1}{2}(\dot{u}^2+\dot{v}^2+\dot{w}^2).$

Thus, the normal coordinates are related to the original coordinates x, y, z by the equations

$$x = \frac{1}{\sqrt{(2m)}}(u+v\sqrt{3}+w\sqrt{3}), \quad y = \frac{1}{\sqrt{(2m)}}(u-w\sqrt{3}),$$

$$z = \frac{1}{\sqrt{(2m)}}(u-\sqrt{3}v+\sqrt{3}w). \quad \square$$

EXERCISES

1. Eight equal point electric charges e are fixed at the vertices $(\pm a, \pm a, \pm a)$ of a cube. Show that the potential energy of another charge e when it is at the point (x, y, z) is given approximately by

$$V = \frac{8e^2}{a\sqrt{3}}\left[1+\frac{7}{162a^4}(3y^2z^2+3z^2x^2+3x^2y^2-x^4-y^4-z^4)\right],$$

when x, y, z are small. Deduce that the ninth charge cannot rest in stable equilibrium at the centre of the cube. If this charge is constrained to move along a diagonal of the cube, show that the centre is a position of stable equilibrium.

2. A uniform bar AB of length $18a$ and mass $6m$ is smoothly hinged at A so that it can rotate in a vertical plane. A smooth ring of mass $3m$ is threaded on the bar and is connected to a point C at a height $16a$ vertically above A by a light elastic string. The modulus of the string is $14mg$ and its

natural length is $8a$. Show that the system has a configuration of equilibrium in which the angle BAC is $60°$ and the ring is distant $6a$ from A, but that the equilibrium is unstable.

3. A uniform triangular lamina OAB can turn freely in a vertical plane about the point O which is fixed. The lamina has mass $3m$ and is isosceles with $OA = OB = 2a$ and the angle AOB is $120°$. A particle of mass m is suspended from A by a string of length a and an equal particle is suspended from B by a string also of length a. Small vibrations take place in the vertical plane OAB. Show that the frequencies of the normal modes are in the ratios $\sqrt{3}:1:2$. If θ is the angle made by AB with the horizontal and ϕ, ψ are the angles made by the strings with the downward vertical, all angles being in the same sense, show that the normal coordinates are given by
$$u_1 = (\tfrac{1}{2}ma^2)^{\frac{1}{2}}(\phi-\psi), \quad u_2 = (\tfrac{1}{6}ma^2)^{\frac{1}{2}}(8\theta+\phi+\psi), \quad u_3 = (\tfrac{1}{3}ma^2)^{\frac{1}{2}}(\theta-\phi-\psi).$$

Index